PHYSICS RESEARCH AND TECHNOLOGY

ATOMIC MULTINOLOGY

PHYSICS RESEARCH AND TECHNOLOGY

Additional books in this series can be found on Nova's website under the Series tab.

Additional E-books in this series can be found on Nova's website under the E-book tab.

PHYSICS RESEARCH AND TECHNOLOGY

ATOMIC MULTINOLOGY

TAEHO WOO
EDITOR

Nova Science Publishers, Inc.
New York

Copyright © 2011 by Nova Science Publishers, Inc.

All rights reserved. No part of this book may be reproduced, stored in a retrieval system or transmitted in any form or by any means: electronic, electrostatic, magnetic, tape, mechanical photocopying, recording or otherwise without the written permission of the Publisher.

For permission to use material from this book please contact us:
Telephone 631-231-7269; Fax 631-231-8175
Web Site: http://www.novapublishers.com

NOTICE TO THE READER

The Publisher has taken reasonable care in the preparation of this book, but makes no expressed or implied warranty of any kind and assumes no responsibility for any errors or omissions. No liability is assumed for incidental or consequential damages in connection with or arising out of information contained in this book. The Publisher shall not be liable for any special, consequential, or exemplary damages resulting, in whole or in part, from the readers' use of, or reliance upon, this material. Any parts of this book based on government reports are so indicated and copyright is claimed for those parts to the extent applicable to compilations of such works.

Independent verification should be sought for any data, advice or recommendations contained in this book. In addition, no responsibility is assumed by the publisher for any injury and/or damage to persons or property arising from any methods, products, instructions, ideas or otherwise contained in this publication.

This publication is designed to provide accurate and authoritative information with regard to the subject matter covered herein. It is sold with the clear understanding that the Publisher is not engaged in rendering legal or any other professional services. If legal or any other expert assistance is required, the services of a competent person should be sought. FROM A DECLARATION OF PARTICIPANTS JOINTLY ADOPTED BY A COMMITTEE OF THE AMERICAN BAR ASSOCIATION AND A COMMITTEE OF PUBLISHERS.

Additional color graphics may be available in the e-book version of this book.

LIBRARY OF CONGRESS CATALOGING-IN-PUBLICATION DATA
Woo, Taeho.
 Atomic multinology / Taeho Woo.
   p. cm.
 Includes index.
 ISBN 978-1-61209-830-2 (softcover)
 1. Nuclear engineering. 2. Technological forecasting. I. Title.
 TK9153.W66 2011   621.48--dc22   2011004576

Published by Nova Science Publishers, Inc. † New York

CONTENTS

Preface		vii
Chapter 1	Introduction	1
Chapter 2	Information Technology (IT)	11
Chapter 3	Biological Technology (BT)	33
Chapter 4	Nano Technology (NT)	61
Chapter 5	Conclusion	97
Acknowledgment		117
Index		119

PREFACE

From the atomic bomb experiment in the mid 20th century, the technological applications in the atomic energy have been studied. The nuclear power plant (NPP) has showed the critical impact on the industry as the aspect of the energy production. Several countries have the big portions in the electrical generation using the atomic power. In France, the rate is over than 80 % in the national electricity production.

However, the political situation has blocked the new NPP construction especially in the United Stated during last 30 years. So, the alternative technology has been developed like the medical physics in the nuclear technology field. This is accompanied with the general trend in the 21st century, which is called IT (Information Technology), BT (Biological Technology), and NT (Nano-scale technology). This book pioneers for the area of the new technology in the atomic industry. Any other works have not focused on the 3 kinds of technological investigations simultaneously. The contents are based on 130 major references which were published separately. These are focusing on the atomic technology as the aspect of interdisciplinary trend.

Most contents are cooperated with the American standards, which mean the new kinds of challenging researches are including. This book could be a guider in the class of the graduate level or senior undergraduate year. Additional references are supplied in the end of chapter. The main object of this book is for the basic knowledge in the areas of three fields. Although this book is for the nuclear engineering technology, the other areas can use as a reference of the research. As the technologies have progressed rapidly, the contents of the book will be extended in near future.

This book is just a basic step for the multi-combinational research promotions. We will really see very great research outputs for the atomic technology field in the 21st century. Those should be contributed to humankind's welfare for a better world.

Taeho Woo
Seoul, Korea
December, 2010

Chapter 1

INTRODUCTION

After a long period's seeking for the letting-out of the inactivity in the global nuclear industry, the new nuclear engineering technology has been developed. This comes up with the interdisciplinary technology of the Information Technology (IT), Biological Technology (BT), and Nano-scale Technology (NT). These areas are perusing for the combinational and the better efficient purpose characteristics. In 21st century, the science and technology are related to be more practical and professional like the conventional business management and medical bases researches.

In the nuclear industry, after Three Miles Island (TMI) nuclear power plant (NPP) accident in 1979, the commercial NPP industry has been suffering from severe stagnation in the marketing business. After another crucial accident of the Chernobyl NPP in 1986, the worldwide nuclear communities have failed to find the alternative industrial promotions. In 1990s, the radiation related researches have been developed in the industry and medicine. But, it has not been successful due to the shortages of industrial demand. In early 2000s, the generation 4 (Gen-4) nuclear power plants (NPPs) have been initiated [1]. But, there was a very unavoidable limitation of the commercial promotion. Those suggested reactor types were the substances that were failed in early stage of nuclear industry during 1950s and 1960s. It, therefore, is necessary for us to make the new technology to support for the Gen-4 initiative. In addition, the Global Nuclear Energy Partnership (GNEP) needs another kind of the aspect for the achievement of the commercialization [2 - 4]. In this philosophy, the GNEP needs the nuclear waste processing and it's related to the fast reactor. These had already developed during the mid and late-20th century, but have not been succeeded yet. So, we need to find the solution of the last century's limitation of the nuclear technology for the industrialization. Only way to make the industrialization is to keep the progress for the general tendency of the nuclear engineering which are related to the IT, BT, and NT.

The algorithm for definition of the Atomic Multinology is seen in Figure 1.1 where the conventional nuclear technology is applied by the IT, BT, and NT. There is the example of the nano-technology case in which the new name is given as Atomic Nanomics. Historically, the nuclear engineering was created for the nuclear power plant manufacturing by several technologies like nuclear physics, mechanical engineering, chemical engineering, nuclear chemistry, and some other interdisciplinary areas in the mid-20th century. This trend is seeking for another creative prospect by interdisciplinary methods. For example, the nano-technology reflected the researches of the atomic research national laboratories in the United States Department of Energy which had a major role in the science and technology during the 20th century. After ending of the cold war, those laboratories had sought for the new research areas by the adaptation to new era. Also some other national research institutes like the National Institute of Health (NIH) and the Environmental Protection Agency (EPA) had joined this new trend of the research.

For the characteristics of this book, the strategy for Atomic Multinology is in Figure 1.2. Namely, the technology is described by the Monte-Carlo (MC) method for the simulations as well as some measurement experiments. Although the experimental results are given, the computational performances are basically shown using the MC method which is most important tool in the atomic technology. The MC is related to all aspects of IT, BT, and NT. However, the experimental measurements are much more focusing on BT and NT than IT.

This book is the fundamental approach for the applications of the atomic technology into the industry using IT, BT, and NT. Later, a new kind of book will be published as the characteristics of encyclopedia. There are lots of researches which have been performed in these areas. The objects of the new researches and developments are surely the better productions of the electricity which is incorporated with a smarter utility in the general industries.

1.1. BASIC THEORY

The Atomic Multinology is fundamentally performed by the MC method [5 - 10] and the solid state-material measurement studies [11 - 26].

Introduction

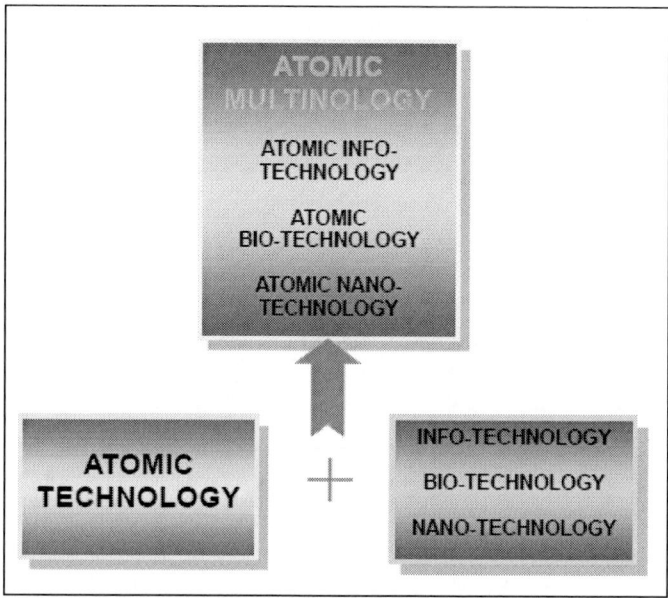

Figure 1.1. Definition of the Atomic Multinology.

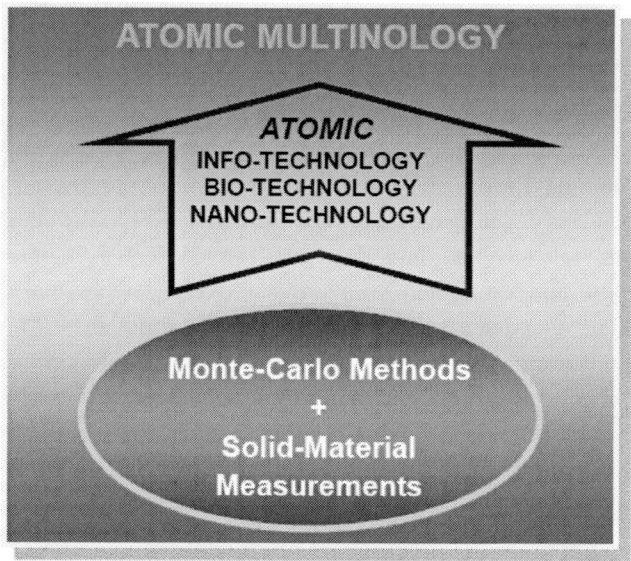

Figure 1.2. Strategy of this book for Atomic Multinology.

It is possible to account for any quantitative analysis and decision making using the MC technique, which is used by professionals in such widely disparate fields as manufacturing, engineering, finance, project management, energy, research and development, insurance, oil-gas, transportation, and the environment. The MC can give the decision-maker wide products of possible and the probabilities for any choice of action. The decisions could be from a variety of cases which will be happened everyday. Historically, this was developed for atomic bomb invention. MC simulation has been used to model for many applications after the introduction in World War II. Name of Monte Carlo is the Monaco resort town renowned for its casinos.

Although the MC is called generally, it is a stochastic techniques-meaning which is based on the use of random numbers and probability statistics to investigate problems. The applications are from physics to management problems, which is one of very important characteristics. The random number generations are used for the specified cases in the problems. Usually, the experiments of scientific as well as social matters are compared with the theoretical simulations by MC method. This way gives the reader of a paper the higher reliability of the modeling. For example, the matter of atomic interactions is a large system which can be sampled in a number of random configurations, and that data can be used to describe the system as a whole.

There are some simulation codes for the MC. Monte Carlo N-Particle Transport (MCNP) and Monte Carlo N-Particle eXpanded (MCNPX) are famous simulation code systems which were developed in Los Alamos National Laboratory (LANL) for simulating nuclear processes. In addition, the Electron Gamma Shower (EGS) computer system for the transport of electrons and photons was developed in Stanford University [27]. Also, there are several applications for business and finance. The System Dynamics (SD) is a methodology and computer simulation modeling technique for framing, understanding, and discussing complex issues and problems [28].

For the experiment, it is necessary to make use of modern nano-structural and nano-chemical analysis techniques including electron microscopy (SEM, TEM, and LEEM), surface microanalysis (SIMS, AES, and XPS), scanning probe microscopy (AFM, STM), x-ray scattering in all modes, and ion-beam spectroscopes (RBS and NRA). The followings are some explanations.

- Scanning Electron Microscope (SEM): The SEM is a microscope that uses electrons rather than light to form an image. Many advantageous analyses are used instead of a light microscope. This has a large depth of field, which allows a large amount of the sample to be in focus at one time. A high resolution images could be produced by SEM, which is that closely spaced features can be examined at a high magnification. It is easy to make the sample due to the sample to be conductive.

- Transmission electron microscopy (TEM): The principle is like the slide projector which produces a beam of light through (transmits) the slide, as the light passes through it is affected by the structures and objects on the slide. Only certain part of the light beam is transmitted through certain parts of the slide due to the above reason. A transmitted beam is then projected onto the viewing screen, forming an enlarged image of the slide.
- Low Energy Electron Microscopy (LEEM): This was developed in 1985 by Ernst Bauer and Wolfgang Telieps, more than 20 years after its invention by E. Bauer in 1962. Otherwise, the first one was installed at the University of Clausthal in Germany. Then, Dr. Ruud Tromp and Dr. Marc Reuter initiated the development of a new one at IBM. It started operation in 1991 and was later sold to several places worldwide. Several kinds of designs have been developed by new theoretical concepts for the correction of imaging errors of magnetic lenses, magnetic deflectors, and analyzers, which are the energy filtered LEEM for chemical analysis, the spin polarized LEEM (SP-LEEM) for magnetic imaging, the Baby LEEM approach and the SMART project. IBM built the first LEEM including a 90° beam deflector worldwide in 1998.
- Secondary Ion Mass Spectroscopy (SIMS): This operates on the principle that bombardment of a material with a beam of ions with high energy (1-30 keV) results in the ejection or sputtering of atoms from the material. The secondary ions are ejected as a small percentage. Any information of the sample compositions is obtained by the collection of these sputtered secondary ions and their analysis by mass-to-charge spectrometry gives information on the composition of the sample. This is used by the locally destructive technique of analyzing material removed from the sample by sputtering
- X-ray photoelectron spectroscopy (XPS): A quantitative spectroscopic technique measures the elemental composition, empirical formula, chemical and electronic state of the elements that exist within a material. Measuring the kinetic energy and number of electrons that escape from the top 1 to 10 nm of the material by X-ray

beam irradiating are shown as the XPS spectra. Ultra high vacuum (UHV) conditions are necessary.

- Atomic force microscopy (AFM): This is a very high-resolution type of scanning probe, with demonstrated resolution on the order of fractions of a nanometer, more than 1,000 times better than the optical diffraction. The precursor to the AFM was developed by Gerd Binnig and Heinrich Rohrer in the early 1980s at IBM Research-Zurich, a development that earned them the Nobel Prize for Physics in 1986. The first commercial one was developed in 1989. This is one of the foremost tools for imaging, measuring, and manipulating matter at the nanoscale. The information is collected by 'feeling' the surface with a mechanical probe. Piezoelectric elements facilitate tiny, but accurate and precise movements on (electronic) command enable the very precise scanning. Electric potentials can also be scanned using conducting cantilevers in some variations. The currents can even be passed through the tip to probe the electrical conductivity or transport of the underlying surface in newer more advanced versions, but this is much more challenging with very few groups reporting reliable data.
- A scanning tunneling microscope (STM): This is an instrument for imaging surfaces at the atomic level. In 1981, the newer more advanced versions were made by Gerd Binnig and Heinrich Rohrer (at IBM Zürich) which got the Nobel Prize for Physics in 1986. A good resolution is considered to be 0.1 nm lateral resolution and 0.01 nm depth resolution for an STM. Usually, STM can be used not only in ultra high vacuum but also in air, water, and various other liquid or gas ambient, and at temperatures ranging from near zero kelvin to a few hundred degrees Celsius.
- Rutherford Backscattering (RBS): Lord Ernest Rutherford presented the concept of atoms having nuclei in 1911. There are a measuring the number and energy of ions in a beam which backscatter after colliding with atoms in the near-surface region of a sample at which the beam has been targeted.
- Neutron Reflection Analysis (NRA): It is known the reflection of neutrons (neutron reflectometry) at a surface is very similar to the phenomenon of the reflection of light (optical reflectometry).
- The reflection of the neutron is like the light which was demonstrated by Fermi. There is a light of strong interference depending on the wavelength of the light, its state of polarization, the thickness of the layer and the refractive indices of the media involved.

Introduction

The nano-scopic simulation's codes for the molecular behavior in the aspect of the solid state and materials are listed as follows [29];

- *Abalone (Molecular dynamics, visualization)*
- *ACEMD (Molecular dynamics)*
- *AMBER (Molecular dynamics)*
- *BOSS (Biochemical and Organic Simulation System)*
- *BRAHMS (Molecular dynamics)*
- *CASTEP (Density-functional theory)*
- *CCP5 (Program Library, various)*
- *CHARMM (Chemistry at HARvard Molecular Mechanics)*
- *CPMD (Molecular dynamics)*
- *Dalton (Computational chemistry)*
- *DiMol2D (Molecular dynamics)*
- *DL_MESO (Dissipative particle dynamics)*
- *DL_POLY (Molecular dynamics)*
- *DYNAMO (Molecular dynamics)*
- *EGO VIII (Molecular dynamics)*
- *ENCAD (Molecular dynamics)*
- *ESPResSo (Molecular dynamics)*
- *FOCUS (Molecular dynamics)*
- *Gaussian (Electronic, Computational chemistry)*
- *gdpc (Molecular dynamics visualization)*
- *GROMACS (Molecular dynamics)*
- *GROMOS (Molecular dynamics)*
- *HOOMD (Molecular dynamics)*
- *IMD (Molecular dynamics)*
- *Jmol (Visualization)*
- *QMGA (Visualization)*
- *RasMol (Visualization)*
- *RedMD (Molecular dynamics)*
- *SageMD (Simulation front and back end)*
- *SIESTA (Molecular dynamics)*
- *SMMP (Monte Carlo simulation)*
- *SYBYL (Various)*
- *Tesla Bio Workbench (GPU computing)*
- *TINKER (Software tools for molecular design)*
- *UHBD (Brownian dynamics)*
- *VASP (Molecular dynamics)*

- *VMD (Visualization, Molecular dynamics visualization in 3-dimensions)*
- *WIEN2K (Electronic structure calculation in solids)*
- *XCrysDen (Visualization, Crystalline and molecular structure visualization)*
- *X-PLOR (Computational structural biology)*
- *YASARA (Fee and commercial)*
- *YASP (Molecular dynamics)*

In addition, Towhee is a Monte Carlo molecular simulation code originally designed for the prediction of fluid phase equilibrium using atom-based force fields and the Gibbs ensemble with particular attention paid to algorithms addressing molecule conformation sampling [30]. The code has subsequently been extended to several ensembles, many different force fields, and solid (or at least porous) phases. Furthermore, SRIM (The Stopping and Range of Ions in Matter) is a collection of software packages which calculate many features of the transport of ions in matter [31].

REFERENCES

[1] U.S.DOE, A Technology Roadmap for Generation IV Nuclear Energy Systems, 2002.
[2] National Economic Council, "Advanced Energy Initiative," February 2006.
[3] The White House, "President Announces New Measures to Counter the Threat of WMD," Fact Sheet, February 2004.
[4] Nuclear Engineering International, "*GNEP is dead; long live Gen-4,*" July 2009.
[5] E.E. Lewis, W.F. Miller, Jr., Computational methods of neutron transport, Wiley, New York, 1984.
[6] István Manno, Introduction to the Monte-Carlo method, Akadémiai Kiadó, Budapest, 1999.
[7] Rubinstein, Reuven Y., Simulation and the monte carlo method, John Wiley and Sons, Hoboken, N.J, 2008.
[8] Ilya M. Sobol, A primer for the Monte Carlo method, CRC Press, Boca Raton, 1994.
[9] http://www.palisade.com/risk/monte_carlo_simulation.asp
[10] http://www.chem.unl.edu/zeng/joy/mclab/mcintro.html

[11] Charles Kittel, Introduction to solid state physics, Wiley, New York, 1986.
[12] J.R. Hook, H.E. Hall, Solid state physics, Wiley, London, 1991.
[13] Giuseppe Grosso, Giuseppe Pastori Parravicini. Solid state physics, Academic Press, San Diego, 2000.
[14] Michael P. Marder, Condensed matter physics, John Wiley, New York, 2000.
[15] Leonard M. Sander, Advanced condensed matter physics, Cambridge University Press, Cambridge, 2009.
[16] Michael F. Ashby and David R.H. Jones, Engineering materials 1 : an introduction to properties, applications and design, Elsevier Butterworth-Heinemann, Amsterdam, 2005.
[17] http://mrl.illinois.edu/facilities.html
[18] http://mse.iastate.edu/microscopy/whatsem.html
[19] http://www.unl.edu/CMRAcfem/temoptic.htm
[20] http://www.research.ibm.com/leem/
[21] http://en.wikipedia.org/wiki/Atomic_force_microscopy
[22] http://en.wikipedia.org/wiki/Scanning_tunneling_microscope
[23] http://www.siliconfareast.com/SIMS.htm
[24] http://en.wikipedia.org/wiki/X-ray_photoelectron_ spectroscopy
[25] http://www.cea.com/training/tutorials/rbs_theory_tutorial
[26] http://rkt.chem.ox/.ac.uk/techniques/nrmain.html
[27] EGSnrc, maintained by the Ionizing Radiation Standards Group, Institute for National Measurement Standards, National Research Council of Canada.
[28] Forrester, Jay W. (1961). Industrial Dynamics. Pegasus Communications.
[29] Wilfred F. van Gunsteren, and Alan E. Mark "Validation of molecular dynamics simulation," Journal of Chemical Physics 108, 6109-6116 (1998)
[30] http://towhee.sourceforge.net/
[31] J. F. Ziegler and J. P. Biersack and M. D. Ziegler (2008). SRIM - The Stopping and Range of Ions in Matter. SRIM Co.

Chapter 2

INFORMATION TECHNOLOGY (IT)

2.1. ABSTRACT

The information technology (IT) is applied to nuclear industry from the nuclear reactor theory to the risk management including the thermal-hydraulics. New kind of non-linear algorithm is introduced for the radiation dispersion for the Chernobyl accident as well as the risk management. Basically, the statistical manipulations for the nuclear interactions are concerned. Even the hardware for the electronics in radiation measurement is classified in the IT field of atomic technology.

2.2. BASIC STATISTICS

One of the important things about any information is how to treat it scientifically. There are many statistical descriptions in the natural phenomenon. The random process is done in dice, radioactive decay, interaction between material and radiation, etc. This is explained as follows;

- Statistical process in measurement
- Error and reliability in the results
- Statistical inspection in the device function

The probability is defined as follows;

- $p(x)$: Probability of event x
- N : Repeat count of observation
- n : Number of event x among N observations

- n/N : Relative frequency if occurrence of x

So, the probability distribution function (pdf) is as follows;

$$p(x) \equiv \lim_{N \to \infty} \frac{n}{N} \qquad (1)$$

The frequency distribution function (fdf) by an estimation of $p(x)$ is,

$$f(x) = \frac{n}{N} \qquad (2)$$

There are two kinds of the random variables. The discrete random variable is x_i, $i = 1, 2, \cdots, s$. So,

$$pdf : p_i \to \sum_{i=1}^{s} p_i = 1 (normalization) \qquad (3)$$

Otherwise, the continuous random variable is x, $a \leq x \leq b$. Then,

$$p(x) \to \int_a^b p(x)dx = 1(normalization) \qquad (4)$$

where $p(x)$ is the probability of x to be in the interval $[x, x + dx]$. The indices of any distribution are described as follows;

- 0th moment (I_0) : normalization
- 1st moment (I_1) : mean or expectation
- 2nd moment (I_2) : variance

Therefore,

$$I_0 \equiv \int_{-\infty}^{+\infty} p(x)dx = 1 \qquad (5)$$

$$I_1 \equiv \int_{-\infty}^{+\infty} x \cdot p(x)dx = \mu (= \overline{x}) \qquad (6)$$

$$I_2 \equiv \int_{-\infty}^{+\infty}(x-\mu)^2 \cdot p(x)dx = \sigma^2 \rightarrow \sigma = \sqrt{I_2} \tag{7}$$

The distributions are described as following three types. The binominal distribution is described by two kinds of cases. If the probability of event A is p, the probability of event B is q (= $1 - p$). For the case of observations of n times, the probability of event A is,

$$P(x,n) = \frac{n!}{x!(n-x)!} p^x (1-p)^{n-x} \tag{8}$$

So,

$$\sum_{x=0}^{n} P(x,n) = 1 \tag{9}$$

$$m = \overline{x} = \sum_{x=0}^{n} x \cdot P(x,n) = np \tag{10}$$

$$\sigma^2 = \overline{(x-m)^2} = \sum_{x=0}^{n}(x-m)^2 \cdot P(x,n) = m(1-p) = np(1-p) \tag{11}$$

If the distribution of binominal distribution with m (= np) is finite and constant, namely,

$$n \rightarrow \infty, \ p \rightarrow 0$$

where,

$$P(x,n) = \frac{n!}{x!(n-x)!} p^x (1-p)^{n-x} \xrightarrow[p \to 0]{n \to \infty} \frac{1}{x!} n^x p^x e^{-m} = \frac{m^x}{x!} e^{-m}$$

$$\frac{n!}{(n-x)!} = n(n-1)\cdots(n-x+1) \xrightarrow{n \to \infty} n^x \tag{12}$$

where,

$$\frac{n!}{(n-x)!} = n(n-1)\cdots(n-x+1) \xrightarrow{n \to \infty} n^x$$

(13)

$$(1-p)^{n-x} = (1-p)^{-x}(1-p)^n = (1-p)^{-x}(1-p)^{m/p} \xrightarrow[p \to 0]{} e^{-m} \qquad (14)$$

Then,

$$\sum_{x=0}^{n} P(x,n) = 1 \qquad (15)$$

$$m = \bar{x} = \sum_{x=0}^{n} x \cdot P(x,n) = np \qquad (16)$$

$$\sigma^2 = \overline{(x-m)^2} = \sum_{x=0}^{n}(x-m)^2 \cdot P(x,n) = m = \bar{x} \qquad (17)$$

If the mean (m) and variance (σ^2) are two independent parameters to describe the Gaussian distribution as $P(x; m, \sigma)$. So, this is shown from passion distribution as follows;

$$P(x,m) = \frac{m^x}{x!} e^{-m} \xrightarrow[m>20]{} \frac{1}{\sqrt{2ps}} \exp\left[-\frac{(x-m)^2}{2s^2}\right] \qquad (18)$$

2.3. COMPUTATIONAL SIMULATION

There are many computational code systems for the NPP. The risk management area is one of them. In recent trend, the non-probabilistic quantification is emerging like the artificial intelligence and the non-linear algorithm. The expert system, fuzzy set theory, neural network theory, and genetic algorithm are designed for the artificial intelligence application. There is another non-linear algorithm which is used as the systems thinking method, simply System Dynamics (SD) [1, 2, 3, 4]. This has produced several commercial softwares as the Stella, Powersim, and Vensim [5]. This kind of method is originated from the business management and the ecological analysis.

The SD was created by Dr. J. Forrest in Massachusetts Institute of Technology (MIT) for the quantifications of the systematic situations. The applications for the non-linear characteristics of the social and economic system have been studied. For the quantification, it is to test and model the complex features in the dynamical scenarios of the interested matters. The feedback of the event and the time step are particular characteristics of the SD, where the event flows are expressed in the non-linear algorithm. The quantification is done by the MC simulations of the defined algorithm.

M. Radzicki described the SD, which is a powerful methodology and computer simulation modeling technique for understanding, framing, and discussing complex issues and problems [6]. It is imagined for managers to improve their understanding, which is practicable in all kinds of policy and design areas. The fundamental block could be expressed by the SD for how and why complex real-world systems behave the way they do during the specified time. The SD can prospect for the understanding to implement much more effective policies. The most important thing is the dynamic behavior of system, where the operator tries to identify the patterns of behavior exhibited by interested system variables, and then builds a model with the characteristics of patterns. In SD molding, the single and double arrow lines are used for the purpose. Usage of lines means the event flows and time flows. The dynamic behavior of a system is manipulated by the dynamic behavior of a system, its key physical and information flows, stocks, and feedback structures for SD. Several characteristics of the SD are explained as follows;

- Nonlinearity: Large part of the SD modeling process involves the application of common sense to dynamic problems. Such behaviors usually indicate a nonlinearity of the events. This is seen as single and double arrow lines in the modeling. Namely, the arrow line shows the event flow without any restriction.
- Stock-flow: The principle of accumulation is performed to be raised by dynamic behavior, which means that all kinds of dynamic behaviors could be happened when flows accumulate in stocks. Both informational and non-informational object can move through flows and accumulations in stocks.
- Feedback: The part of feedback loops is shown by the stocks and flows in real world systems. The feedback loops are often joined together by nonlinear couplings where any object often causes counterintuitive behavior.
- Time Paths: The dynamic behavior of systems is quantified, where the operator tries to identify the patterns of behavior exhibited by interested system variables, and then builds a model with the characteristics of patterns.

There are special expressions for the above characteristics in the SD modeling. Especially, in the Vensim code, the technical methods are done by single and double arrow lines as follows;

- Single arrow line: This line shows the flow of the event, which means it is the sequence of the scenarios as well as the dynamical behavior. So, the direction of line gives the event flow and event feedback.
- Double arrow line: The dynamic behavior is raised in the SD modeling for the principle of accumulation. All kinds of dynamic behaviors could be happened when flows accumulate in stocks, which are seen as EXAMPLE for accumulation and INPUT/OUTPUT for flows in Figure 2.1. Figure 2.2 shows the causal loop of the Figure 2.1 where the event flow is seen. This is like a bathtub where a flow can be thought of a faucet and pipe assembly that fills or drains the stock. It is thought as the simplest dynamical system in the stock-flow structure. Both informational and non-informational object can move through flows and accumulation in stocks for the SD. It is thought that the feedback loops are often joined together by nonlinear couplings where any object often cause counterintuitive behavior, which is seen as curved loop in Figure 2.1. A plus sign means for the addition to EXAMPLE of the feedback value, OUTPUT. Otherwise, if the sign is minus, the feedback value, OUTPUT, is subtracted from the EXAMPLE.

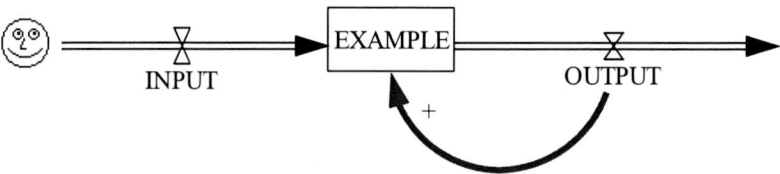

Figure 2.1. Stock- flow and Feedback.

Figure 2.2. Causal loop.

There are some publications for the organizations by the transitions of the time [7, 8, 9, 10]. In addition, there are also the decision-making related papers [11, 12, 13]. The dynamic simulation methods using the SD are commercial software packages as the Vensim [14], Powersim [15], and ITHINK [16] which are applied for the quantifications. For this chapter, the Vensim is used for the simulations.

The non-linear algorithm also has applied for the nuclear reactor theory and the thermal-hydraulics. Although this kind of logic has not been used in the commercial nuclear industry area, there are many research works which have been performed. There is a difficulty to apply a new technology to a real situation due to the conservativeness of a certain industrial field. In addition, the regulation should be modified to new types. However, the conventional linear calculation needs the non-linear ways due to its limitation of the solution finding.

The failure frequency is changed by time step. The time step is changed following the situation of NPP like standby-running and running-shutdown. The high failure frequency rate is able to be shown in the case of shutdown for refueling or trip. The basic events are weighted by time step which means the time step is a feedback factor expressing the situation of NPP. The dynamics approach can show the human factor using operator's time dependent situation. The common cause failure is made by time step process. The modeling is easily made by graphic designed method. The figures are understood by operator or reviewer well. The availability and capacity are calculated through the simulation. In the conventional Probabilistic Risk Assessment (PSA), this was just found by operation data.

2.4. RADIOISOTOPE DISPERSION

The Cesium-137 dispersion is modeled for the methodological dynamic simulation in Figure 2.3. This is a critical issue in the NPP accident like Chernobyl case which happened in the late Union of Soviet Socialist Republics (USSR). The lake and soil are considered for the time dependent radioisotope modeling. This modeling is affected by feedback factors like temperature, wind velocity, wind direction, and so on. This simulation is compared with the real measured data. The radioisotope containing air flows as the turbulent manner in the troposphere and stratosphere. The dispersion of Cesium-137 is calculated by time step which expresses the time of the isotope's travel in the designed area. The air quantity like surface wind speed is affected by some variables which are modified in the condition of wind strength by random sampling. The data can make new modeling for the radioisotope dispersion model equation. Using the formula, one can expect the new presumable accident scenario and the relevant variables can be changed by the estimator's decision and other external data.

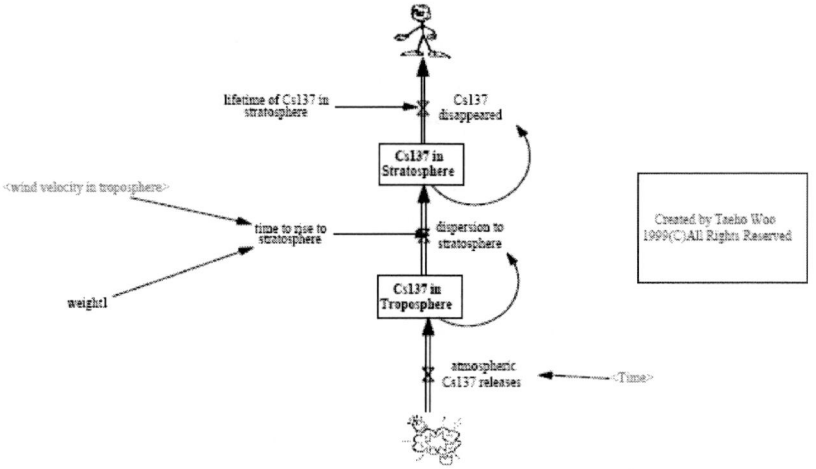

Figure 2.3. Cs-137 dispersion by Chernobyl accident.

This can be applied to the chlorofluorocarbon (CFC) case in ozone layer and population of specified animals or plants are well estimated for the designated dynamical result.

The sensitivity analysis is shown in Figure 2.4. Each region is changed as seen in the figure.

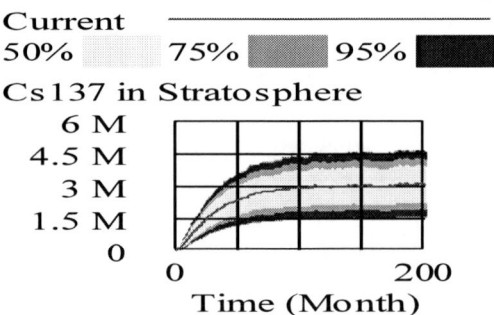

Figure 2.4. Continued.

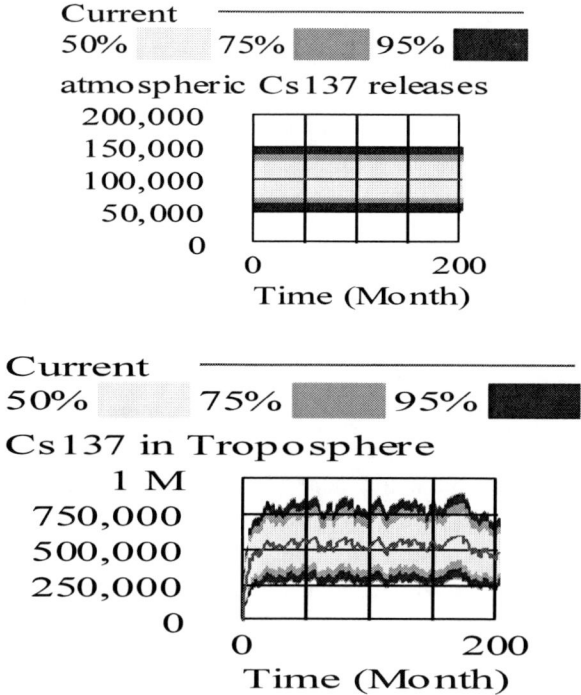

Figure 2.4. Sensitivity analysis of Cs-137 dispersion by Chernobyl accident.

2.5. SAFETY MANAGEMENT

2.5.1. Background

The safety management is performed for the nuclear power system as safety assessment. The non-linear dynamical safety assessment is introduced for the analysis of the High Temperature Gas Cooled Reactor (HTGR) which is applicable for the Modular Pebble Bed Reactor (MPBR). The dynamical algorithm is adjusted for the safety assessment. An easier and reliable output is addressed. The poison xenon is contributed to the decrease of reactor power. The operator's action is investigated for the manual action of safety control rod axe man (SCRAM). The feedback of power increase causes to the temperature decrease. The accident scenario is changeable following the proposed conditions. The realistic analysis is constructed in the simple algorithm.

The top event of the event is studied as Power and Temperature Stable. It is not always completed successfully in the designed periods. Namely, it is affected by the human factor, poison, and some other physical variables. So, this work manifests that the accident scenario cannot guarantee the safety measure situation in the very danger accident like the gas duct failure. Each analyzed step of the sequence of the accident shows much more reliable results using relevant tools like the System Dynamics algorithm.

2.5.2. Method and Result

The HTGR is a very efficient NPP for the energy demand, especially in the current era of the unstable oil energy market situation. It is suggested for the competitive cost and safety aspects comparing to the coal and oil energy sources. The hydrogen production is one of main objects for the HTGR. Some NPPs have been constructed and operated for the pre-step of the commercialization. So, the accident modeling is important and it should be done as the aspects of advanced algorithm like the non-linear complex system for the real situation.

In the HTGR, the DAWS could be developed for the core damage involvement. The reactor transient code is developed and compared with conventional method for the DAWS [17]. The Severe accident Free HTR (SFHTR) is investigated, which has been developed for the power with 450MW$_t$ in Japan Atomic Energy Research Institute (JAERI) [18, 19, 20]. It is necessary to assess the new model. Simpler and more exact calculation is developed using feedback logic of the systems in the safety assessment.

An advanced dynamical simulation algorithm in NPP is developed using SD algorithm. In 1950s, the SD was developed for the logic of feedback control to the interested system in the electrical and control systems [7]. The Systems Thinking has its basis in the SD. This helps people to make their understanding of social systems clear and enhance them in the same way that people can use engineering principles to make clear and enhance their understanding of mechanical systems. The commercial software for SD, Ventana Simulation Environment (Vensim), is used. The quantification of the assessment in the accident is simply and easily calculated comparing to the conventional method due to the visual graphics like the arrow mark, causal diagrams, and other figures.

The scenario of depressurization in the gas duct of HTGR is investigated. Each step is physically connected each other. However, there is still the possibility of the malfunction of the series of steps. So, the physical reaction and human action should be considered that is an important point for the sequence. The scenario of the accident is done in the Figure 2.5. This is modified from the original procedure of Nakagawa [17]. The feedback algorithm shown in the power increase makes the temperature decrease due to the negative reactivity effect of the reactor. The real situation is affected by several factors like the human factor, feedback nature of the system, and other physical variables. The event success is noted as 1 and the failure is 0. The output is described as graphs in each step of the events. The each event is quantized as random number calculations. That is to say, the conventional probability is changed to the operator defined random number generation.

DEPRESSURIZATION ACCIDENT SAFETY ASSESSMENT OF HTGR

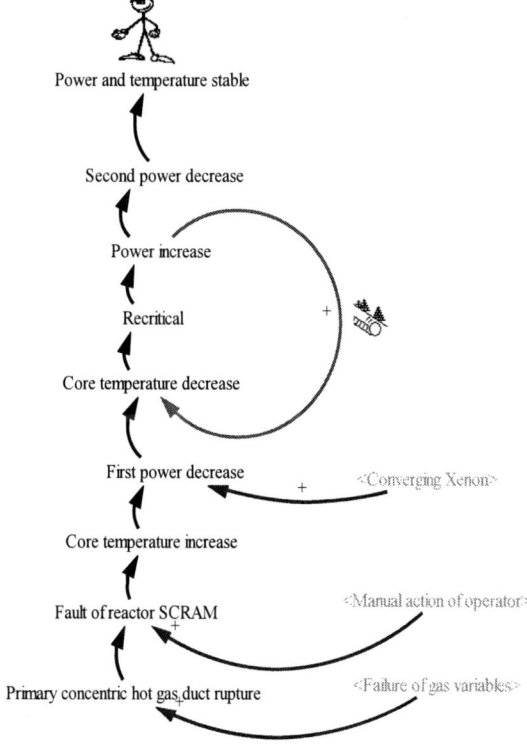

Figure 2.5. Event sequence of DAWS.

The temperature and reactivity are shown to be stabilized in the most points during the 200 hour period. The First power decrease and the Second power decrease are caused by the negative reactivity effect. Following the temperature decrease, the reactivity is stabilized. This is to be a Recritical result which is in the feedback loop.

It is affected by the feedback nature of the system by the reactivity due to the characteristics of nuclear reactions. The human factor of the operator causes to the SCRAM action. The operators are working by the working cycle which is assumed as 8 hours which makes 3 times change sequences during a day.

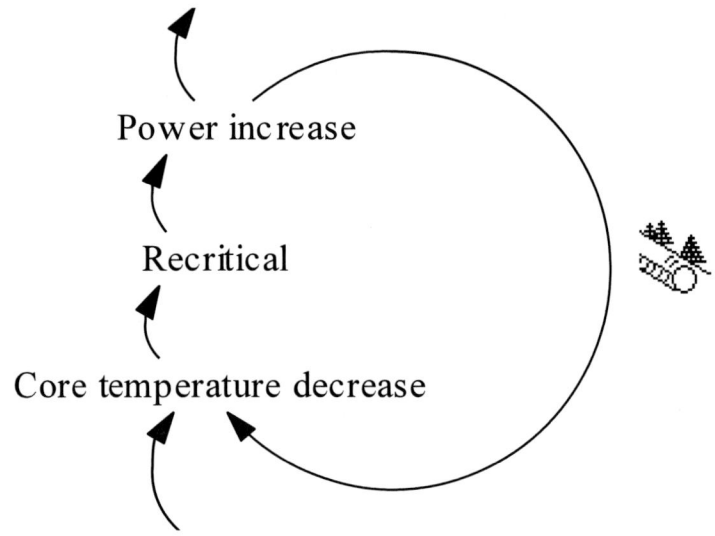

Figure 2.6. Feedback loop of Recritical.

Figure 2.6 shows the feedback loop of the Recritical where the Power increase is feedback to Core temperature decrease. The relationship between work cycle and Manual action of operator incorporated with Time function is in Figure 2.7. The operator has the 8 hour working period in Figure 2.8, because it is the usual working cycle in the site. So, the performance of the work is changed by the Work Cycle in Figure 2.8. In the middle of cycle (around 4 hour), the working efficiency is assumed to be highest. The relationship of Converging Xenon loop is shown in Figure 2.9. There is the Gas variable failure loop in Figure 2.10. Xenon is a role of the poison in the reactor. The quantity is changed by the half-life and decay factors. The pressure and temperature are main factors in the failure of the gas coolant for the system. The final results show the 38 times failures of the event in the power and temperature stable, which is in Figure 2.11. This result value is one of the operator's conditions in this paper and should be very low in the real plant operation.

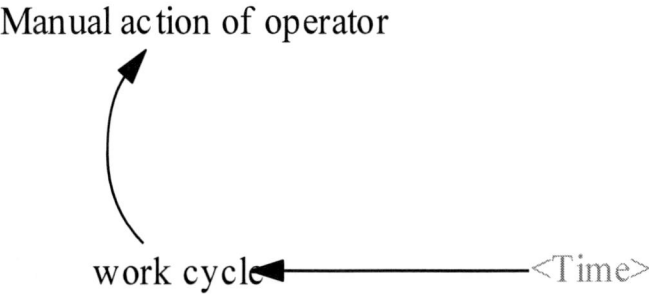

Figure 2.7. Manual action of operator loop.

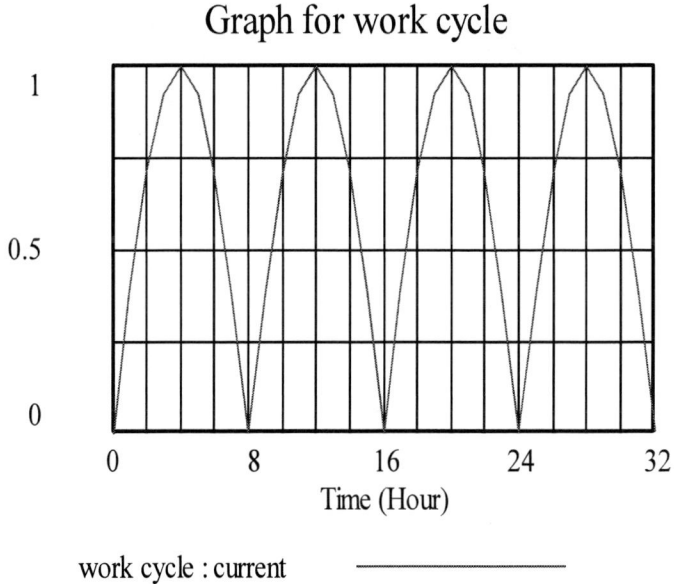

Figure 2.8. Work cycle graph.

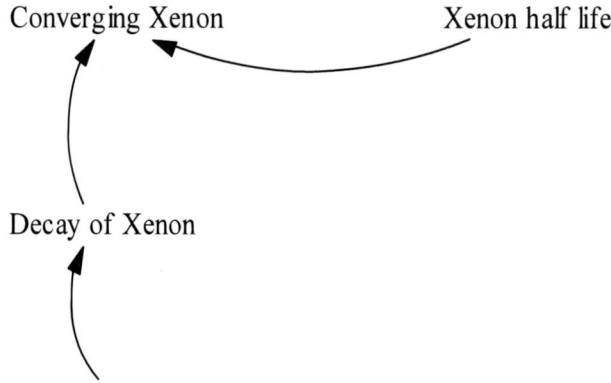

Figure 2.9 Converging Xenon loop.

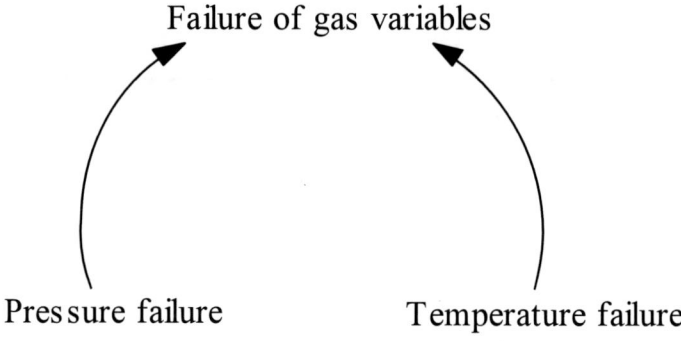

Figure 2.10. Gas variable failure loop.

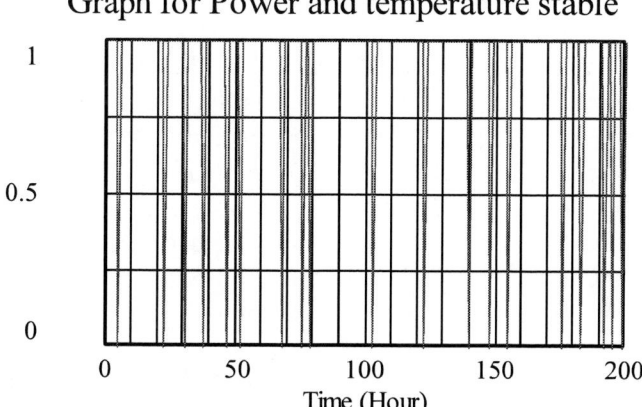

Figure 2.11. The power and temperature stable graph.

2.5.3. Summary

The comments for the conclusion of this simulation are as following explanations. The failure of the temperature and reactivity stable are done frequently during the Time Step of the 200 hours. The poison xenon affects to the each step of the scenario. Criticality is reproduced by the negative characteristics. Human factor is one of main factors to adjust the 8 hours' working period during 200 hours cycle.

The top event of the event is power and temperature stable which is in Figure 2.11. It is not always successful in all periods. It is affected by the human factor, poison, and some other physical variables. The proposed accident in the NPP can produce unexpected results. So, it is necessary to analyze each step of the sequence using relevant tools like the SD algorithm. There are several limitations in the description for the accidents like the human factor and time interval in the conventional ways. In this paper, those variables are easily described by the feedback algorithm and the Time Step concepts. The non-linear feedback oriented safety analysis is obtained.

REFERENCES

[1] T. H. Woo, S. M. Kwak, 'Human-System Interface Study using System Dynamics Aspects for the Control Room Operator', Nuclear Installations Safety Division International Meeting on Advanced Nuclear Installations Safety, San Diego, June 5-6, 2000.

[2] T. H. Woo, S. M. Kwak, 'The Advanced Logical Dynamics Safety Assessment using System Dynamics Method of Auxiliary Feed Water System (AFWS)', Nuclear Installations Safety Division International Meeting on Advanced Nuclear Installations Safety, San Diego, June 5-6, 2000.

[3] T. H. Woo, S. M. Kwak, 'Dynamical Assessment of Radioisotope Atmospheric Dispersion Using System Dynamics Approach in Caesium-137', The conference record of AGU 1999 Fall Meeting, San Francisco, CA, Dec. 13-17, 1999.

[4] Sterman, J. D., 2000. Business Dynamics : Systems thinking and modeling for a complex world. McGraw Hill.

[5] Ventana Systems, Inc., 2009. Vensim software.

[6] Radzicki M, Taylor R, U.S. Department of Energy's Introduction to System Dynamics, A Systems Approach to Understanding Complex Policy Issues, 1997;1.

[7] Forrester, J., 1961. Industrial Dynamics, Pegasus Communications, Waltham, MA.

[8] Forrester JW, Principles of Systems, 2nd Ed. Pegasus Communications, 1968.

[9] Forrester JW, Urban Dynamics. Pegasus Communications, 1969.

[10] Forrester JW, World Dynamics. Wright-Allen Press, 1971.

[11] Forrester JW, Collected Papers of Jay W. Forrester. Pegasus Communications, 1975.

[12] Kampmann, CE, Feedback loop gains and system behavior. Proceedings of the1996 International System Dynamics Conference Boston. System Dynamics Society, Albany, NY, 1996;260-263.

[13] Liehr M, Grobler A, Klein M, Milling PM, Cycles in the Sky : Understanding and Managing Business Cycles in the Airline Market, System Dynamics Review 2001;17(4):311-332.

[14] Vensim, Ventana System, Inc.

[15] PowerSim, Powersim Software.

[16] ITHINK Software, ISEE Systems, Inc.

[17] Nakagawa, S., et al., 1989. JAERI-M 89-013.

[18] Kunitomi, K. et al, 1996. Proceeding of the JSME/ASME 4th nt'l Conf. On Nucl. Eng. 1996. Vol. 2, pp 303 – 307.
[19] Satto, S., et al, 1994. JAERI 1332.
[20] Kunitomi, K., et al., 1989. JAERI-M 89-001.

PROBLEMS

1. Solve the following example.

Tom and Jane decide to assign grades at random to the students in the medical imaging class. Jane calls out the name of a student while Tom rolls a dice. If the dice shows a value of 1, then the student get an A. If the dice shows a different value, then the student gets a B. If there are 20 students in the class, what is the probability that,

a) Only one student gets A and B in the class
b) 3 or fewer students gets an A in the class

Solution

a) For this question, betting a 'B' is considered a 'success' with a probability of P = 5/6.
   For n = 20, the probability of $x = 1$ students receiving a B is,

$$P(1, 20) = \frac{n!}{x!(n-x)!} p^x (1-p)^{n-x} = \frac{20!}{1!(20-1)!} \left(\frac{5}{6}\right)^1 \left(1-\frac{5}{6}\right)^{20-1} = 20 \left(\frac{5}{6}\right)^1 \left(\frac{1}{6}\right)^{20-1}$$
$$= 2.735 \times 10^{-14}$$

b) For this question, betting an 'A' is considered a 'success' with a probability of P = 1/6.
   For n = 20, the probability of $x = 0$ students receiving an A is,

$$P(0,20) = \frac{20!}{0!(20-0)!}\left(\frac{1}{6}\right)^0\left(1-\frac{1}{6}\right)^{20} = \left(\frac{5}{6}\right)^{20} = 0.0261$$

$$P(1,20) = \frac{20!}{1!(20-0)!}\left(\frac{1}{6}\right)^0\left(1-\frac{1}{6}\right)^{20-1} = 20\left(\frac{1}{6}\right)\left(\frac{5}{6}\right)^{19} = 0.1043$$

$$P(2,20) = \frac{20!}{2!(20-2)!}\left(\frac{1}{6}\right)^2\left(1-\frac{1}{6}\right)^{20-2} = \frac{20 \cdot 19}{2}\left(\frac{1}{6}\right)^2\left(\frac{5}{6}\right)^{18} = 0.1982$$

$$P(3,20) = \frac{20!}{3!(20-3)!}\left(\frac{1}{6}\right)^3\left(1-\frac{1}{6}\right)^{20-3} = \frac{20 \cdot 19 \cdot 18}{2}\left(\frac{1}{6}\right)^3\left(\frac{5}{6}\right)^{17} = 0.2379$$

The probability of getting at 3 or fewer students receiving A's in the class is,

$$P(0,20) + P(1,20) + P(2,20) + P(3,20) = 0.5665$$

2. You perform a counting experiment with a weak radioisotope source and Geiger counter in which you measure 269 counts in minute

(a) Derive the Poisson probability distribution which describes the probability of obtaining x counts in 1 second.
(b) Assuming that the radioisotope source has a relatively long half-life (so that the count rate dose not change during you measurement), what is the probability of measuring 3 or more counters in any 1 second interval of time?

Solution

(a) The mean number of counters in 1 second m is,

$$m = \frac{269}{60} = 4.48[counts/sec]$$

$$P(x, m = 4.48) = \frac{e^{-m}m^x}{x!} = \frac{e^{-4.48}4.48^x}{x!}$$

(b) The probability of measuring 3 or more counters/sec is,

$$P(x, m = 4.48) = \frac{e^{-m} m^x}{x!} = \frac{e^{-4.48} 4.48^x}{x!}$$

$$P(x \geq 3) = 1 - P(x \leq 2) = 1 - [P(0) + P(1) + P(2)]$$
$$= 1 - 0.01133 - 0.05077 - 0.113733 = 0.824$$

$$P(0) = \frac{e^{-4.48} 4.48^0}{0!} = \frac{e^{-4.48}}{1} = 0.01133$$

$$P(1) = \frac{e^{-4.48} 4.48^1}{1!} = 0.05077$$

$$P(2) = \frac{e^{-4.48} 4.48^2}{2!} = 0.113733$$

3. You've assigned a student helper to make measurements on a radio labeled protein and instructed him to repeat each measurement 10 times so that you can calculate the mean and standard deviation of the measurement. The student works on this project and when he returns. He reports that he has made 10 measurements and obtained the following values : 105, 104, 107, 102, 101, 106, 105, 107, 107, 108.

(a) Calculate the mean and standard deviation of this measurement, and show that the measured standard deviation is smaller than that you would expect from a counting experiment.
(b) Is the observation from (a) possible for a radiation counting experiment? Discuss possible reasons why the measured standard deviation could be smaller than the theoretically predicted value?

Solution

(a) $$s_x = \sqrt{\frac{\sum_{i=1}^{N}(x_i-\bar{x})^2}{N-1}} = \sqrt{\frac{\sum_{i=1}^{10}(x_i-105.63)^2}{10-1}} = 2.62$$

$$\therefore \sigma_x^2 = (2.62)^2 = 6.85$$

For a Poisson-distributed event, we should obtain $\sigma^2 = m = 105.63$. Thus, the data as given is not from a Poisson-distributed event.

(b) The experimental observation is not possible from a counting experiment. There must be an instrumentation error or the student is recording the data incorrectly.

4. Discuss fuzzy set theory.

Solution

It is explained the fuzzy sets are sets where elements have degrees of membership. The fuzzy sets were created by Lotfi A. Zadeh (1965) as an extension of the classical notion of set. The elements in a set are assessed in binary terms according to a bivalent condition in classical set theory. This means that an element belongs or does not belong to the set. Otherwise, the fuzzy set theory allows the conditional existence of the membership of elements in a set. Namely, this is described with by a membership function which is in the unit interval [0, 1].

5. What is system dynamics (SD)?

Solution

System dynamics (SD) was introduced by Professor Jay Forrester of the Massachusetts Institute of Technology (MIT) during the mid-1950s. In 1956, as a professor in MIT Sloan school of management, he determines how his background in science and engineering could manipulation the core issues that determine the success or failure of corporation for the useful way. The SD is a methodology and computer simulation modeling technique for framing, understanding, and discussing complex issues and problems. This helps the managers improve their understanding of industrial processes where the SD is currently being used throughout the public and private sector for policy analysis and design.

Chapter 3

BIOLOGICAL TECHNOLOGY (BT)

3.1. ABSTRACT

The biological technology is applied to medical area by the nuclear energy. The radiation is used usually for the radiation diagnostics and therapy. Especially, the cancer therapy is an important issue in the clinical radiation treatment. The radiological diagnostics for the cancer are much used in clinical treatments like the mammography which is used for the breast cancer diagnostics. The radiological therapy is the usage of the radiation energy for the cancer cells. The brachytherapy is commonly used for the treatment of the cervical cancer. Ion therapy makes use of the bragg-peak where the dose is maximized in the organ.

3.2. BASIC CONCEPT

Radiation is used for the biological and medical applications. The decay rate of radiation described as the decay rate per second as the following explanation. If radioactive material quantity is $N(t)$ and decay constant λ, the decay rate $A(t)$ is,

$$A(t) = \frac{dN(t)}{dt} = l \cdot N(t) \tag{1}$$

If one uses,
$$X \equiv \frac{dQ}{dm_{air}}$$
So,
$$N(t=0) = N_o$$
$$A(t=0) = A_o = l\, N_o \tag{2}$$

Then, using half-life $T_{1/2}$,

$$N(t) = N_o e^{-lt} = N_o \left(\frac{1}{2}\right)^{1/T_{1/2}} \tag{3}$$

$$A(t=0) = l \cdot N(t) = l \cdot N_o e^{-lt} = A_o e^{-lt} = A_o \left(\frac{1}{2}\right)^{1/T_{1/2}}$$

The exposure means the concept of a field intensity of radiation field. This is total charge dQ which is produced when the secondary electron by photon stops. The unit is roentgen (X). That is to say,

$$X \equiv \frac{dQ}{dm_{air}} \tag{4}$$

So,

$$X = 1[C/kg_{air}] = 3876R$$
$$1R = 2.58 \times 10^{-4}[C/kg_{air}] \tag{5}$$

The exposure rate is,

$$\dot{X} \equiv \frac{dX}{dt}\left[\frac{C}{kg_{air}}, \frac{R}{s}\right] \tag{6}$$

The absorbed dose (*D*) is the energy in the radiated material as follows;

$$D \equiv \frac{dE}{dm_{matter}} \left[\frac{J}{kg_{air}} \right] \qquad (7)$$

The unit is,

$$1Gy = 1 \left[\frac{J}{kg_{air}} \right] = 100 rad \qquad (8)$$

$$1 rad = 100 \left[\frac{erg}{g_{matter}} \right] = 100 \left[\frac{10^{-7}}{10^{-3} kg_{matter}} \right] = 0.01 [Gy] \qquad (9)$$

So, the absorbed dose rate is,

$$\dot{D} \equiv \frac{dD}{dt} \left[\frac{Gy}{min}, \frac{Gy}{hr}, \frac{mGy}{hr}, \frac{rad}{s}, \frac{rad}{min} \right] \qquad (10)$$

The KERMA is defined as the Kinetic Energy Released in MAterial. The summations of initial kinetic energy of all kinds of charged particles are considered which is produced by non-charged radiations like *x*-ray or *r*-ray.

$$K = \frac{dE_{tr}}{dm_{matter}} \left[\frac{J}{kg_{matter}} \right] \qquad (11)$$

The kerma rate is,

$$\dot{K} = \frac{dK}{dt} \left[\frac{Gy}{s}, \frac{rad}{s} \right] \qquad (12)$$

Otherwise, the CERMA is also defined as the Kinetic Energy Released in MAterial. The summations of initial kinetic energy of all kinds of Charged particles are considered which is produced by charged radiations except *x*-ray or *r*-ray. The CERMA rate is,

$$\dot{C} = \frac{dC}{dt}\left[\frac{Gy}{s}, \frac{rad}{s}\right] \quad (13)$$

The dose equivalent is defined as follows;

$$H = Q \cdot D \quad (14)$$

where, H is dose equivalent, D is absorbed dose, and Q is quality factor. Q is changed by the radiation field.

$$1 Sv = 1\left[\frac{J}{kg_{tissue}}\right] \quad (15)$$

$$1 rem = 100\left[\frac{erg}{g_{tissue}}\right] \quad (16)$$

$$1 Sv = 100 rem \quad (17)$$

The effective does (E) means the summation of dose equivalent multiplied by organ weighting (W_T).

$$E \equiv \sum_T H_T \cdot W_T = \sum_T (Q \cdot D_T) \cdot W_T \quad (18)$$

3.3. RADIATION DIAGNOSTICS

3.3.1. Background

The common radiation diagnostics are to use the chest x-ray in the hospital, which are used frequently in the clinical purposes. One of examples in radiation diagnostics is the mammography for the breast cancer diagnostics. The breast cancer has the highly ranked death rate among all cancer death rates in American women. The early diagnosis is very important in order to decrease the death rate. The best way to decrease mortality is to detect the cancer cell in early stage. U.S. national survival statistics show that the five-year survival rate for breast cancer is 85 % when the cancer is localized to the breast. Otherwise the rate is down to the 56 % when the auxiliary nodes are included. The classical screen/film mammography has been considered as a reliable diagnostic way. However, this calculation causes masses and micro-calculations to be difficult to image the dense breasts. So, the direct digital x-ray acquisition is necessary.

3.3.2. Mammography

The digital mammography is exampled in this book for the explanation for the clinical usage [1]. The Active Matrix Flat-Panel Imagers (AMFPIs) are being developed for x-ray detection systems. Indirect detection imagers typically use $Gd_2O_2S:Tb$ or $CsI:Tl$ scintillation screens to convert the x-ray into visible photons which are then collected by an underlying photodetector array for digital radiographic and mammographic applications. This chapter has been investigating whether or not the inclusion of a microlens array between the screen and photodetector may improve light collection when the photodetector has a small optical fill factor. In this method, the technique for modeling the modulation transfer function (MTF) from measurement obtained for $Gd_2O_2S:Tb$ and $CsI:Tl$ scintillation screens is reported. The measurements were obtained for a number of different mono and polychromatic x-ray (energy) spectra. The screen MTFs were subsequently transformed into point spread functions (PSFs) and used in a simulation of the proposed imaging system. This imaging system makes a better image in the lower radiation exposure to patients.

3.3.3. Object

The milestone of the project is to make a better image in mammography using microlens. Firstly, a good x-ray imager detector as the AMFPI [2], [3] photodetector is constructed. In this stage, the improved ray-tracing [4], [5] and Monte-Carlo simulation are done. The screen, microlens array, and photodetector planes are also considered. Secondly, the microlens array is designed. The transmittance, focusing, and uniformity are measured. The predicted and measured light collection efficiencies are compared. Thirdly, the microlens array is matched to AMFPI. Finally, the comparisons with and without the microlens are studied for the uniform field and phantom images. The simulations of photon detection for two geometrically defined cases in microlens attached photodetector are investigated.

3.3.4. Method

The objectiveness for the investigation is to develop an x-ray detector design including microlens optimized for the AMFPI photodetector. In order to develop the microlens arrays, it is necessary to measure the transmittance, focusing, and uniformity of the microlens arrays [6], [7]. The simplified configuration for this simulation is in Figure 3.1. Figure 3.2 shows the light dispersion by the screens where the molecular structures and the light distributions [8]. The MTF is obtained by experimental data [9]-[17]. MTFs are formulated by the exponential functions which are in Table 3.1. The MTF is converted to Line Spread Function (LSF) using Inverse Fourier Transform [18] and this LSF is also converted to Point Spread Function (PSF) using Inverse Abel Transform in Figure 3.3. Table 3.2 shows the full width at half maximum (FWHM) of PSF. Using theses properties, the light collection efficiency is predicted. The geometries for the calculations [19] are classified as Lambertian and the Isotropic methods. These methods are to simulate how the digital imager configuration makes the light collections. The Lambertian quanta is profiled by,

$$E_{Lamb}(\theta,\phi) = \begin{pmatrix} E_o, & 0 \leq \theta \leq \frac{\pi}{2} \\ 0, & otherwise \end{pmatrix}$$

In addition, the isotropic case,

$$E_{Iso}(\theta,\phi) = E_1 \cdot \sin\theta$$

The θ and ϕ are the meridian and azimuthal angles of the photons.

3.3.5. Result

The microlens diameter is 25 μm lens systems. The distance between a screen and microlens is 5 μm. The light collection efficiency is given as the percentage of the light collected in photodetector. Each point P stands for the pixel point of the photodetector. The P9 is the center pixel and there are pixels from P1 to P8 around P9. Table 3.3 shows each geometrical simulation in Lambertian source and Isotropic source assumptions. The light collection efficiencies for Gd_2O_2S and CsI are shown in this Table. It is shown that the 27 kVp, 82 mm Gd_2O_2S is highest light collection case in both Lambertian and Isotropic geometries. In CsI, 20 keV, 150 mm case has the highest light collection efficiency. For the clinical consideration, the energy range is usually between 17 kVp and 25 kVp. For the next step of experiment, the highest efficiency can be a standard comparison data.

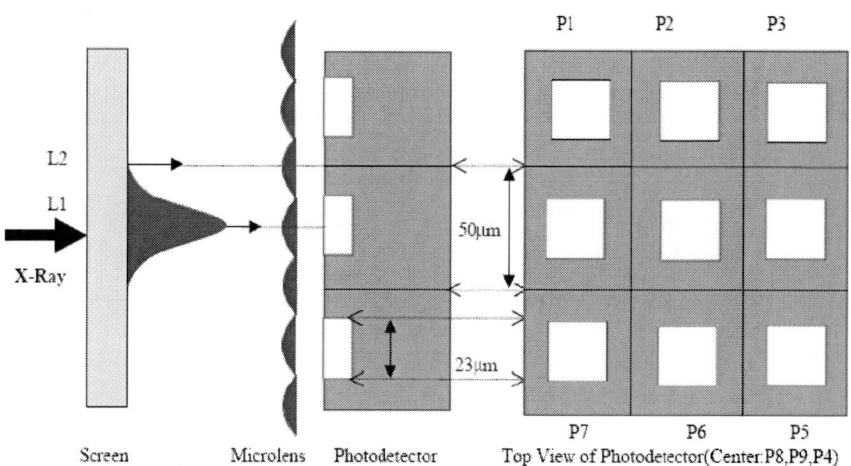

Figure 3.1. Imager configuration.

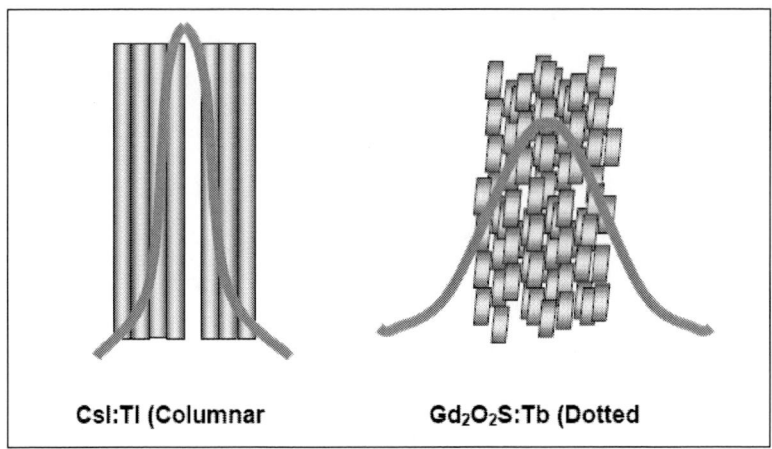

Figure 3.2. Molecular structure and light distribution between CsI and Gd_2O_2S phosphors.

Table 3.1. Phosphor data list.

(a) Gd_2O_2S:Tb

Energy (kVp)	Thickness (μm;mg/cm^2)	FWHM (μm)	MTF (Exponential Function)
25	48; 35	20.96	y=0.5084 exp(-0.087800χ)+0.5476exp(-0.087755χ)
26	46; 34	50.54	y=0.5246 exp(-0.196500χ)+0.5331exp(-0.196400χ)
27	46; 34	42.60	y=0.5954 exp(-0.166400χ)+0.5251exp(-0.166400χ)
	82; 60	92.08	y=0.5160 exp(-0.373300χ)+0.4996exp(-0.372800χ)
28	40; 29	36.81	y=0.5613 exp(-0.153900χ)+0.5337exp(-0.154000χ)
30	44; 32	27.40	y=0.4926 exp(-0.084820χ)+0.5579exp(-0.084490χ)
	68; 50	37.58	y=0.5111 exp(-0.153400χ)+0.5568exp(-0.153700χ)
90	68; 50	38.31	y=0.5117 exp(-0.157000χ)+0.5212exp(-0.156800χ)
	109; 80	51.48	y=0.4813 exp(-0.212198χ)+0.5314exp(-0.212043χ)

(b) CsI:Tl

Energy (kVp)	Thickness (μm;mg/cm^2)	FWHM (μm)	MTF (Exponential Function)
[a]20	150; 68	24.98	y=0.0995 exp(-0.652700χ)+0.9001exp(-0.094440χ)
[b]35	150; 68	29.79	y=0.3429 exp(-0.577000χ)+0.6571exp(-0.105000χ)
[c]50	150; 68	30.68	y=0.2790 exp(-0.535900χ)+0.7207exp(-0.096730χ)
30	150; 68	38.03	y=0.5691 exp(-1.011000χ)+0.4307exp(-0.119900χ)
200	147; 47	29.64	y=0.4726 exp(-0.121500χ)+0.5069exp(-0.121400χ)
	299; 135	41.38	y=0.1424 exp(-65.75000χ)+0.8576exp(-0.143400χ)

[a]20, [b]35, [c]50 are unit of *keV*.

MTF $y = a \exp(-bx) + c \exp(-dx)$

↓   Inverse Fourier Transform

LSF $y = 2ab/(4\pi^2 x^2 + b^2) + 2cd/(4\pi^2 x^2 + d^2)$

↓   Inverse Abel Transform

PSF $f(r^2) = ab/[4\pi^2 \{r^2 + (b/2\pi)^2\}^{3/2}] + cd/[4\pi^2 \{r^2 + (c/2\pi)^2\}^{3/2}]$

Figure 3.3. Transition for point spread function.

Table 3.2. Full with at half maximum (FWHM) list

(a) $Gd_2O_2S:Tb$

Energy (kVp)	Thickness (μm; mg/cm^2)	FWHM (μm)
25	48; 35	21.41
26	46; 34	47.93
27	46; 34	40.60
	82; 60	91.01
28	40; 29	37.56
30	44; 32	20.65
	68; 50	37.46
90	68; 50	38.28
	109; 80	51.75

(b) CsI:Tl

Energy (kVp)	Thickness (μm; mg/cm^2)	FWHM (μm)
[a]20	150; 68	23.09
[b]35	150; 68	25.99
[c]50	150; 68	23.85
30	150; 68	29.73
200	147; 47	29.63
	299; 135	34.98

[a]20, [b]35, [c]50 are unit of *keV*.

Table 3.3 Light collection efficiency list.

(a) $Gd_2O_2S:Tb$

Type	Energy (kVp)	Thickness (μm;mg/cm^2)	L1		L2	
			P9	P1 - P8	P6 - P9	P1 - P5
Lambertian Source (%)	25	48; 35	32.03	0.094	0.013	0.002
	26	46; 34	14.43	0.042	0.209	0.030
	27	46; 34	17.07	0.049	0.174	0.025
		82; 60	76.00	0.220	0.210	0.020
	28	40; 29	18.41	0.053	0.153	0.022
	30	44; 32	33.50	0.097	0.001	0.000
		68; 50	18.46	0.053	0.152	0.022
	90	68; 50	18.07	0.052	0.158	0.023
		109; 80	13.36	0.039	0.220	0.031
Isotropic Source (%)	25	48; 35	11.05	1.105	0.012	0.005
	26	46; 34	4.935	0.493	0.189	0.080
	27	46; 34	5.839	0.584	0.158	0.066
		82; 60	26.00	0.260	0.190	0.080
	28	40; 29	6.300	0.630	0.138	0.058
	30	44; 32	11.46	1.146	0.001	0.000
		68; 50	6.317	0.632	0.137	0.058
	90	68; 50	6.181	0.618	0.143	0.060
		109; 80	4.572	0.457	0.199	0.083

(b) CsI:Tl

Type	Energy (kVp)	Thickness (μm;mg/cm^2)	L1		L2	
			P9	P1 - P8	P6 - P9	P1 – P5
Lambertian Source (%)	[a]20	150; 68	76.00	0.220	0.210	0.030
	[b]35	150; 68	67.52	0.195	0.419	0.060
	[c]50	150; 68	73.58	0.213	0.259	0.037
	30	150; 68	59.25	0.171	0.749	0.107
	200	147; 47	59.23	0.171	0.740	0.106
		299; 135	50.17	0.145	1.2111	0.173
Isotropic Source (%)	[a]20	150; 68	26.00	0.260	0.190	0.030
	[b]35	150; 68	23.10	2.310	0.189	0.160
	[c]50	150; 68	25.17	2.517	0.234	0.099
	30	150; 68	20.19	2.019	0.678	0.285
	200	147; 47	20.26	2.026	0.669	0.282
		299; 135	17.16	1.716	1.096	0.461

[a]20, [b]35, [c]50 are unit of *keV*.

3.3.6. Summary

The light collection efficiency can be increased by the microlense optically focusing method. This depends on the angular distribution of the photons. Originally, the quantum number is by the x-ray energy and screen thickness. As a result, the image blur can be decreased by the photon detection numbers in the photodetector.

3.4. RADIATION THERAPY

The radiation is also used for the disease cell destruction using the radiation energy. The profit of the treatment is that the damage of the normal cell is minimized comparing to the surgery. There are three examples. One of them is the motion control investigation in the patient body where the breath induced movement disturbs the exact radiation irradiation. So it is necessary for a doctor to make the patient's tiny movement stop. The second example is about the Brachytherapy for the HDR (High Dose Rate) and LDR (Low Dose Rate). The last thing is the ion beam therapy. The purpose of this treatment is to minimize the damage of the normal cell due to the radiation exposure using the Bragg's peak principle.

3.4.1. Motion Control

3.4.1.1. Background
The minimization of radiation exposure to the patient is very important in the clinical radiation therapy [20, 21]. One achievement of the purpose is to control the breath. It can be considered by the compensation for the motion of tumor during the dose delivery, because the positions of tumor cell are changeable following the breath of the patient. The computerized software simulation achieves the patient monitoring system, when the x-ray is working on the treatment. The advanced and safe radiation therapy algorithm is constructed [22, 23, 24, 25, 26].

3.4.1.2 Method and Result
The radiation therapy facility at Samsung Medical Center (SMC) in the Republic of Korea is used. The phantom and computer software by LabVIEW are used for investigation of the tumor motion and tracking. The object is to make the algorithm for the patient control in the clinical situation. To achieve the optimized system, there are 3 major stages for the useful protocol construction.

3.4.1.2.1. Record and Analyzing System for Tumor Motion

Signals from a simulator are split and amplified before transmitted to a personal computer. Radiographic images from a simulator are recorded. One or two selected points in a tumor are tracked and analyzed. Figure 3.4 shows the description, where the record and analyzing system for tumor motion is shown.

3.4.1.2.2. Tracking System of the Patient Breathing Motion

A marker is attached to the thorax of a patient. The motion of the marker is recorded in videotape and tracked using a motion tracking software by LabVIEW. Figure 3.5 shows the racking system of the patient breathing motion.

3.4.1.2.3. Tumor Motion Simulating Phantom

A phantom simulating a moving tumor is designed and manufactured. One small metal ball is mounted on a thin bar which is connected to motors and programmed to perform a periodic motion. The description shows the situation of the breathing of the patient when the tumor moves periodically.

The experiments which used the phantom, developed real-time motion compensating system and moving board showed that the deviation is changed from 2.3 cm to 1.2 to 1.3 cm, in X and Y directions, respectively. Tumor motion simulating phantom is shown in Figure 3.6.

3.4.1.3. Summary

There are 3 significant conclusions as follows;

1. The developed system is capable of investigating and analyzing tumor and patient breath cycle.
2. The correlation between the tumor trajectory and the breathing cycle can be investigated using the developed system.
3. The quantitative monitoring system will integrated into a motion adaptive radiotherapy system in the future work.

The real-time motion compensation is achieved using the phantom. However, it is necessary to improve the speed of the program and the control of couch motion as well as this chapter on potential problems that could be caused by couch motion for the clinical application.

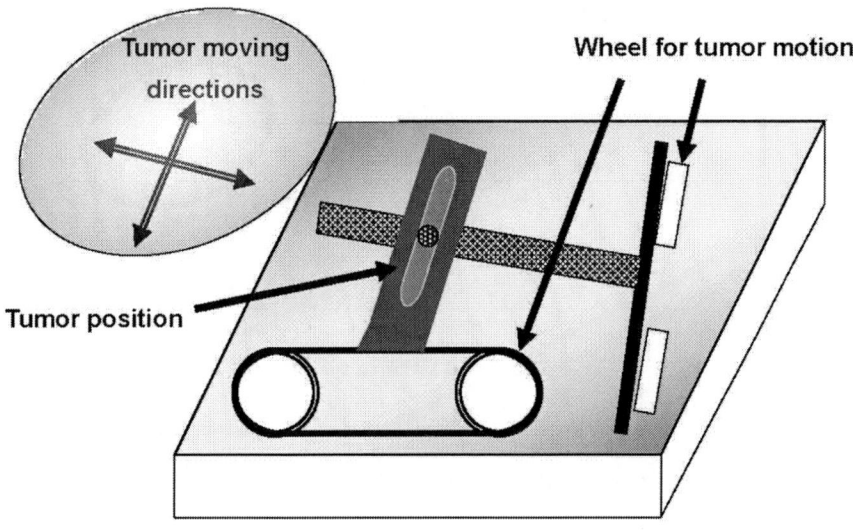

Figure 3.4. Record and analyzing system for tumor motion.

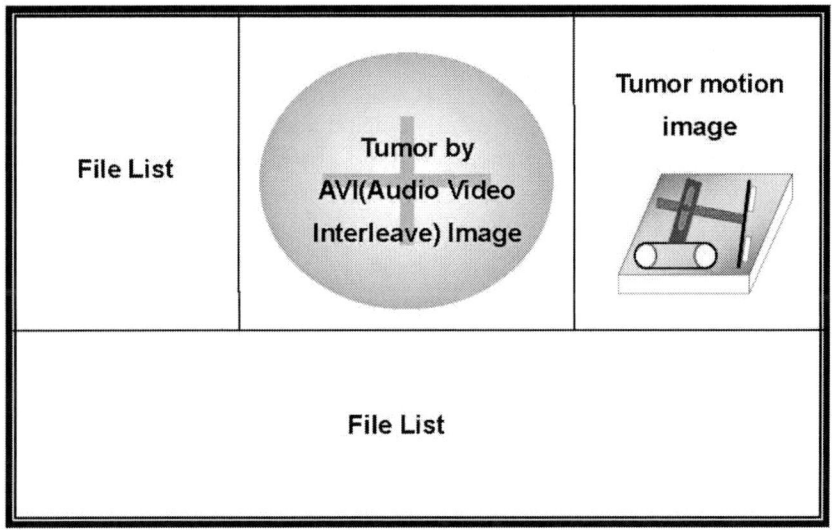

Figure 3.5. Tracking system of the patient breathing motion (Configuration of LabView window).

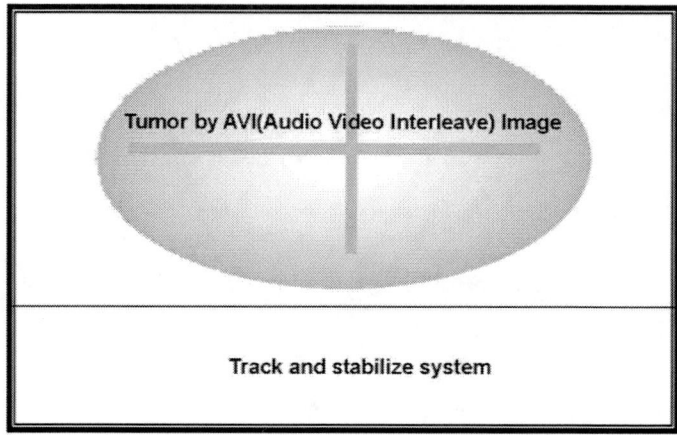

Figure 3.6. Tumor motion simulating phantom (Configuration of LabView window).

3.4.2. Brachytherapy

3.4.2.1 Background

The High Dose Rate (HDR) Brachytherapy protocol has been developed [27]. The purpose of this project is the establishment of safety and accuracy standards of HDR treatment system, which includes the HDR afterloader, source, and radiotherapy treatment planning system (RTP). Many functional and structural features, such as structure of a source, source dwelling method, safety interlock feature and dose calculation accuracy, are unique for each model. It is necessary for the different features of the HDR treatment system to be evaluated by a standardized criterion in order to be approved for the patient treatment [28, 29, 30, 31, 32, 33].

To achieve the described purpose, it is important to calculate for the patient therapy in the clinical applications. Using commercial software, the automatic calculator is developed. It is quietly exact in the RTP. The clinical result is matched with the calculations.

3.4.2.2. Method and Result

The radiation therapy is done by the Nucletron Microelectronic Classic and RTP is Nucletron Plato. The zero point of body coordinate has the one source and the exposure time is 1 minute. There are 8 points for the calculation of the absorption calculation from P1 to P8 as; P0(0, 2 cm, 0), P1(-2cm, 2 cm,

0), P2(2 cm, -2 cm, 0), P3(0, 2 cm, 2 cm), P4(0, 2 cm, 0), P5(0, 5 cm, 0), P6(2cm, 4 cm, 0), P7 (-2 cm, 4 cm, 0), and P8(0, 4 cm, 2 cm).

The dosimeters of interested sources are calculated in the designed geometry. First of all, one needs to decide the source as a point source or cylindrical source in order to find out the geometry factor. According to the Technical Guide (TG) 41, point source is described as $1/r^2$ and line source is $\beta/(L \cdot r \cdot \sin\theta)$. Figure 3.7 shows the line source case. The result of the calculated doses and RTP doses is in Table 3.4.

$$D(r,\theta) = \frac{\Lambda \cdot S_K \cdot g(r) \cdot \phi_{an} \cdot 60}{r^2}$$

Variable	Meaning
r	Dist. bet. source and detection (cm)
θ	Angle of detection point from y axis
$D(r, \theta)$	Absorption dose rate in detection point
Λ	Dose rate constant
S_K	Air-kerma strength
$1/r^2$	Geometry factor, point source approximation
$g(r)$	Radial dose function.
ϕ_{an}	Average anisotropy factor (x axis point=1.0)

S_K= 7.2274 $cGy\ cm^2$ / sec
Λ= 1.12 $cGy / h\ U$

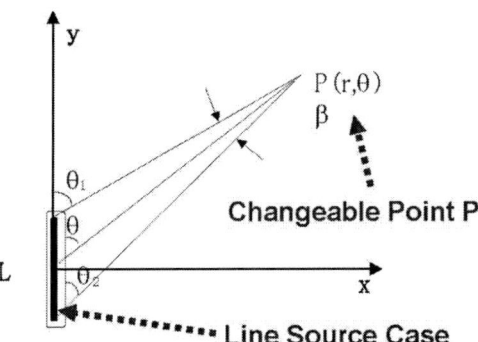

Figure 3.7. The system of line source case.

Table 3.4. Result of the Calculated Doses and RTP Doses

Var.	r	θ	g(r)	φ$_{an}$	Cal. D(r, θ)	RTP D(r,θ)	Dif. (%)
P0	1.9990	0.0000	1.0105	0.6450	79.2162	92.4	-14.27
P1	2.8277	45.0371	1.0149	0.9607	59.2186	59.3	-0.14
P2	2.8277	44.9629	1.0149	0.9607	59.2185	56.7	4.33
P3	2.8277	45.0372	1.0149	0.9606	59.2185	59.0	0.37
P4	1.9990	0.0000	1.0105	0.6450	79.2162	88.4	-10.39
P5	4.9990	0.0000	0.9956	0.6960	13.4665	14.6	-7.76
P6	4.4721	26.5843	1.0047	0.8947	21.8361	22.1	-1.19
P7	4.4721	26.5843	1.0047	0.8947	21.8361	22.2	-1.64
P8	4.4721	26.5843	1.0047	0.8947	21.8361	22.1	-1.19

3.4.3. Ion Therapy

3.4.3.1. Background

The proton is used to treat the cancer cell in the organ. The nano-scale ion is investigated for the behavior in the interaction between the ion beam and the cancer cell [34]. The optimized dose in the interested position of the organ is classified following several variables like beam energy, cell properties, cancer cell depth, and so on. The successful treatment is expected in the clinical results.

The Bragg-peak is examined following the several variables. There is a significant importance for the computational simulations before the construction of the facility due to the huge size of the facility and the very high cost for the installment.

3.4.3.2. Method

The Bragg-peak is examined by the physical property and the ion beam interactions. There are several kinds of radiations are investigated. Higher mass ion has higher dose value in Figure 3.8 [35, 36]. That is to say, the carbon ion has the effective doses between 7 and 8 (arb. Units). Otherwise the proton has it between 5 and 6 (arb. Units). Table 3.5 shows the Organ Compound.

The proton interaction with the human organ is simulated with the computational code systems, SRIM (the Stopping Power and Range of Ions in Matter) [37]. This includes easy and tractable calculations which produce output data of stopping powers, range and straggling distributions for any ion at any energy in any elemental target. Many elaborate calculations include objective targets with complex multi-layer configurations. The proton, which is one of ion beams, is widely used in medical therapy, especially in the area of the radiation oncology. The organs are pancreas and thyroids for the interaction with ion beams. Other organs could be investigated

3.4.3.3. Result

The graphs show the distributions following the energy change. Figure 3.9 shows the energy loss. The energy losses decrease as the proton energy increase. Figure 3.10 shows energy to recoils of the proton in the pancreas. The energy to recoils increases as the proton energy increases. Figure 3.11 shows energy to recoils for the phonons. The energy to recoils decreases as the proton energy increases. The energy loss and energy to recoils increase. Otherwise, in the phonon case, the energy to recoils decreases.

The Figure 3.12 shows the energy losses in the pancreas and thyroid. The pancreas has the larger loss than the thyroid. The Figure 3.13 shows the neon ion has the larger energy losses than the proton. This is useful to treat in the tumor depth by the higher dose of the heavier ion. There are the energy recoils in the figures 3.14, 3.15, and 3.16. There are no consistencies in the reaction between the ion and the organs. The Neon ion has the higher energy recoils, which is similar to the case of the ionization.

3.4.3.4. Summary

The graphs show the interested ion-matter reactions have particular characteristics. Although, the dose is important in the radiation therapy, the characteristics can be changed by the energy of the ions. So, it is important to take the knowledge for the relationship between the energy and the ions.

Table 3.5. Organ Compound

	Pancreas	Thyroid
Chemical Compound	H-11, C-17, N-2, O-69, P-0.4, S-0.4, Cl-0.2, K-0.2	H-10, C-12, N-2, O-7, Na-0.2, Cl-0.2, K-0.2, I-0.1

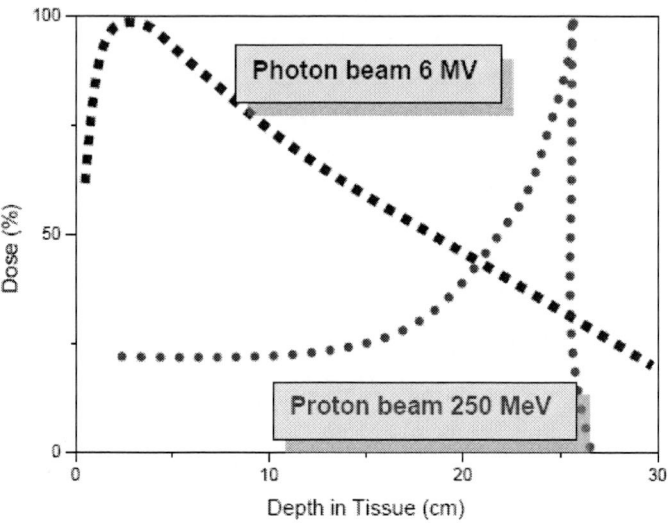

Figure 3.8 Bragg peaks of several cases.

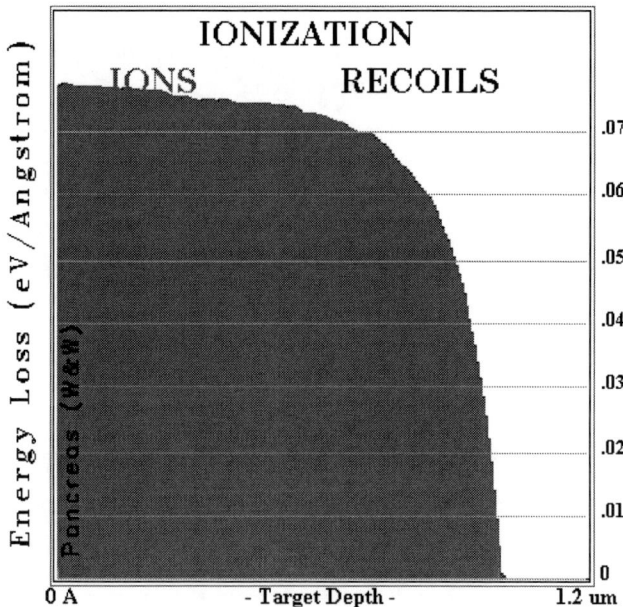

Figure 3.9. Energy Losses (H+ 100MeV).

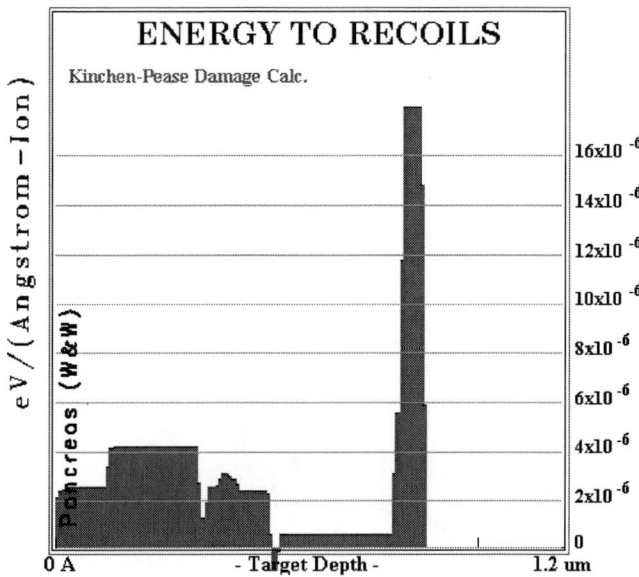

Figure 3.10. Energy to Recoils (H+ 100MeV).

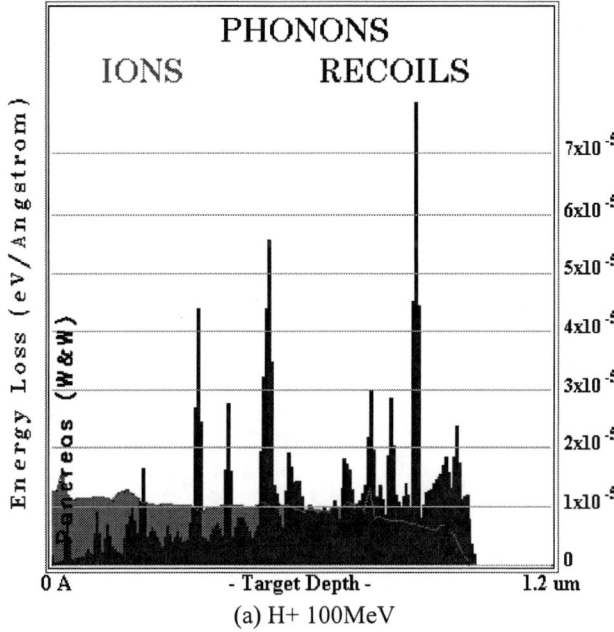

Figure 3.11. Continued.

52 Taeho Woo

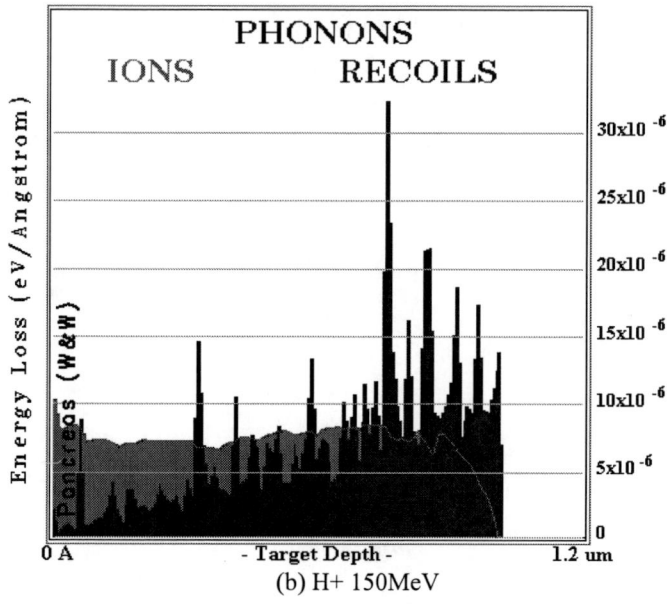

(b) H+ 150MeV

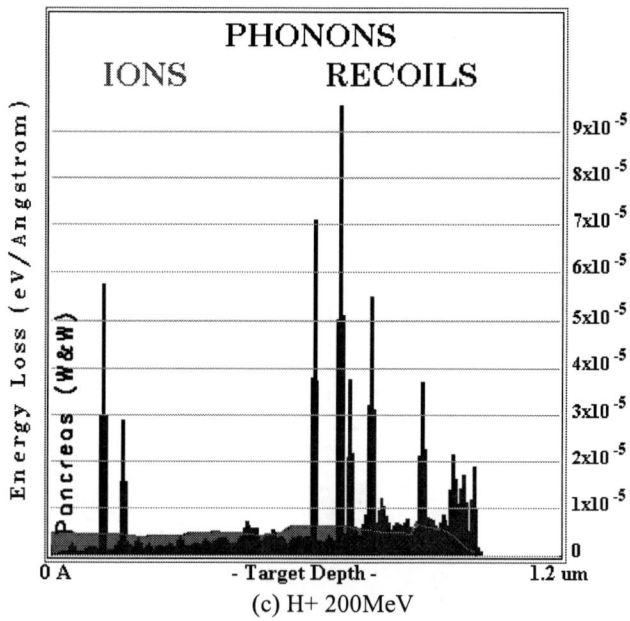

(c) H+ 200MeV

Figure 3.11. Energy to Recoils.

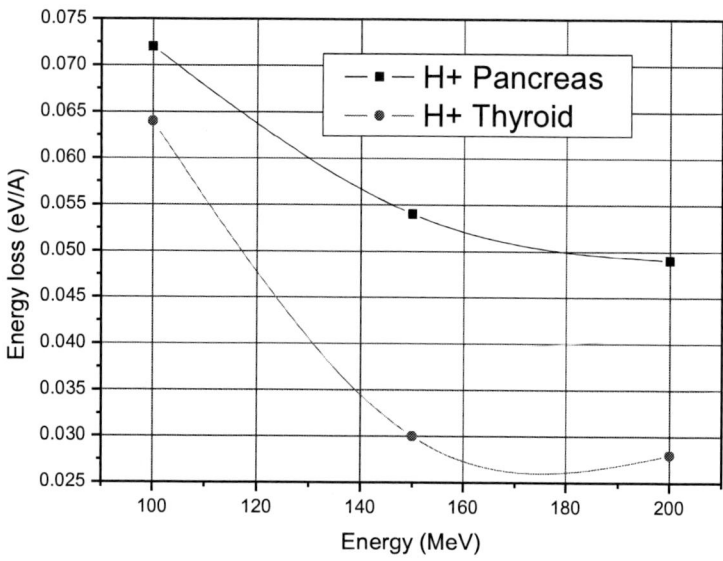

Figure 3.12. Ionizations in Organs (at 600 nm depth).

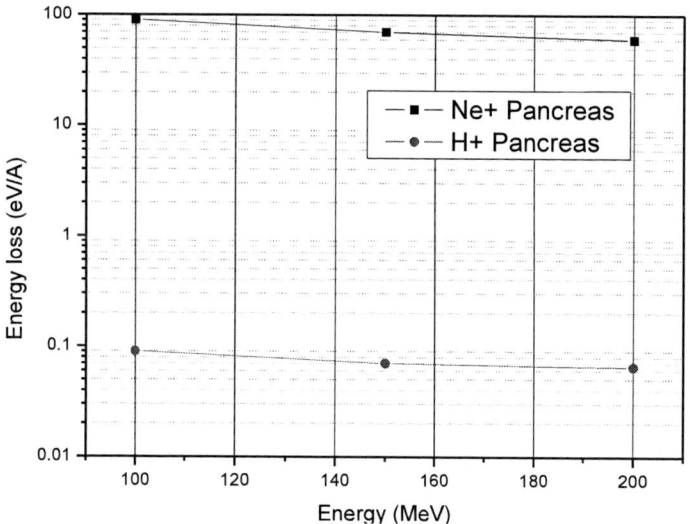

Figure 3.13. Ionizations in Pancreas (at 600 nm depth).

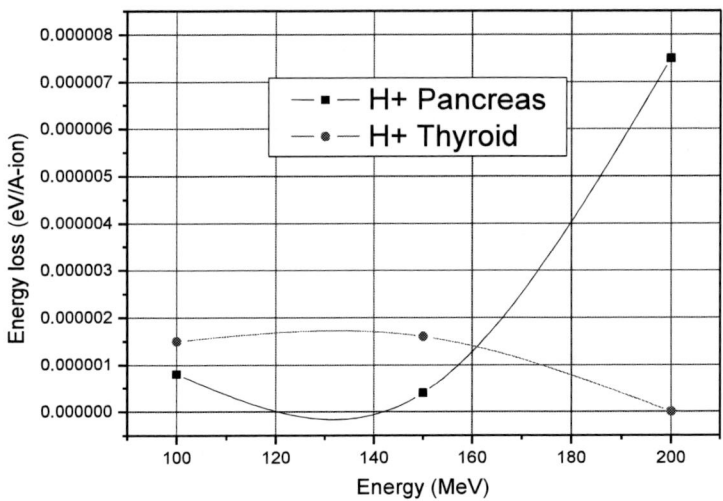

Figure 3.14. Energy to Recoils in Organs (H+) (at 600 nm depth).

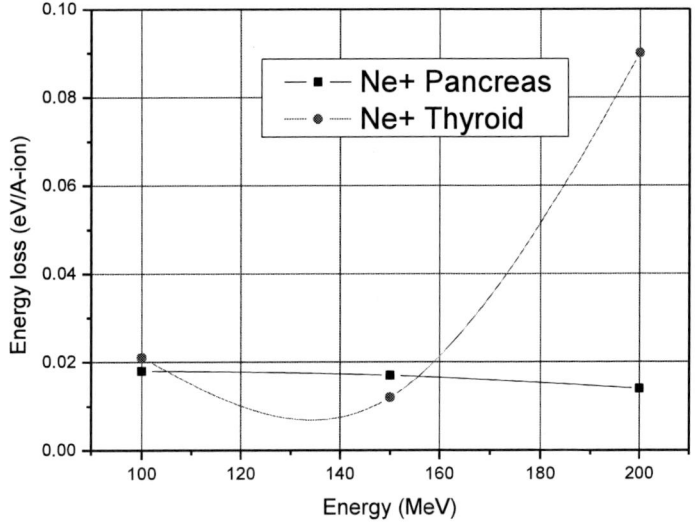

Figure 3.15. Energy to Recoils in Organs (Ne+) (at 600 nm depth).

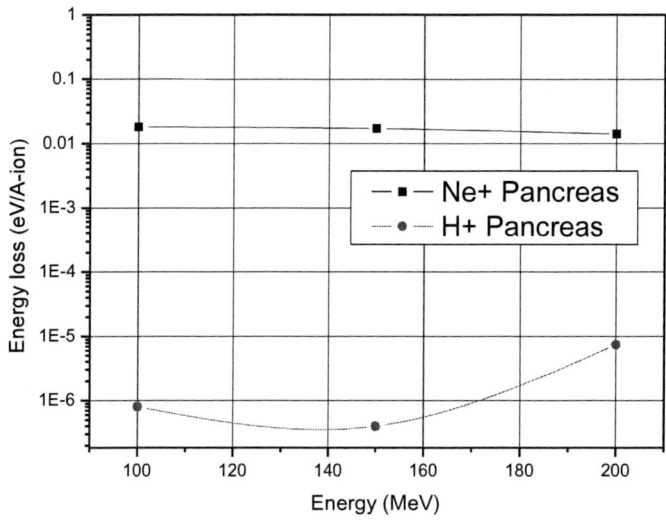

Figure 3.16. Energy to Recoils in Pancreas (at 600 nm depth).

REFERENCES

[1] Nicholas Petrick, Tae-Ho Woo, Heang-Ping Chan, Berkman Sahiner, 'Microlens focusing for enhanced light collection in digital x-ray detectors', *Radiology*, 221(P), 691, 2001.
[2] L. E. Antonuk, Y. El-Mohri, W. Huang, K. W. Jee, M. Maolinbay, R. A. Street, 1997. 'Development of a high resolution, active matrix, flat-panel imager with enhanced fill factor', SPIE, Physics in Medical Imaging, 3032, 3-13.
[3] L. E. Antonuk, Y. El-Mohri, J. H. Siewerden, J. Yorkston, 1997. 'Empirical investigation of the signal performance of a high-resolution, indirect detection, active matrix falt-panel imager (AMFPI) for fluoroscopic and radiographic operations', Medical Physics 24, 512-70.
[4] P. Shirley, 2000. Realistic Ray Tracing, Natick, Mass.
[5] S. Glassner, 1989. An Introduction to Ray Tracing, Academic Press, London.
[6] Nicholas Petrick, Tae-Ho Woo, Heang-Ping Chan, Berkman Sahiner, 2001. 'Microlens focusing for enhanced light collection in digital x-ray detectors', RSNA (Radiological Society of North America) 2001, 87th

Scientific Assembly and Annual Meeting, McCormick Place, Chicago, IL, Nov. 25-30.
[7] Petrick N, Woo TH, Chan HP, Sahiner B, Hadjiiski LM, 2002. 'Evaluation of light collection in digital indirect detection x-ray imagers: Monte carlo simulations with a more realistic phosphor screen model', 6th IWDM (International Workshop on Digital Mammography) 2002, Bremen, Germany, June 22-25.
[8] HAMAMATSU, Inc., Fiber optic plate with scintillator for digital x-ray imaging, 1996.
[9] I. Kandarakis, D. Cavouras, G. S. Panayiotakis, C. D. Nomicos, 1997. 'Evaluating X-ray detectors for radiographic applications : A comparison of ZnSCdS:Ag with Gd2O2S:Tb and Y2O2S:Tb screens', *Phys. Med. Biol.* 42, 1351.
[10] J. Sabol, J. Boon, 1997. SPIE Vol. 3032, 270.
[11] T. Yu, J. M. Sabol, J. A. Seibert, J. M. Boone, 1997. 'Scintillating fiber optic screens : A comparison of MTF, light conversion efficiency, and emission angle with Gd2O2S:Tb screens', *Medical Physics,* 24, 279.
[12] D. Cavouras, et al., 1998. *Appl. Radiat. Isot.,* Vol. 49, No.8, 931-933.
[13] P. Bunch, 1997. Company Internal Data, Kodak company.
[14] I. Kandarakis, D. Cavouras, G. S. Panayiotakis, D. Triantis, C. D. Nomicos, 1997. 'An experimental method for the determination of spatial-frequency-dependent detective quantum efficiency (DQE) of scintillators used in X-ray imaging detectors', Nuc. Inst. Meas. A, 399, 335-342.
[15] Z. Jing, et al., 1999. SPIE Vol. 3659, 163.
[16] V. V. Nagarkar, T. K. Gupta, S. R. Miller, Y. Klugerman, M.R. Squillante, G. Entine, 1998. 'Structured CsI(Tl) scintillators for X-ray imaging applications', *IEEE Trans. on Nuc. Sci.,* Vol. 45, No. 3, 492.
[17] V. V. Nagarkar, J. S. Gordon, S. Vasile, P. Gothoskar, F. Hopkins, 1996. 'High resolution X-ray sensor for non-destructive evaluation', *IEEE Trans. on Nuc. Sci., Vol.* 43, No. 3, 1559.
[18] R. N. Bracewell, 2000. The Fourier Transform and Its Applications, McGraw Hill, Boston.
[19] A. D. Maidment, M. J. Yaffe, 1995. 'Analysis of signal propagation in optically coupled detectors for digital mammography : I. Phosphor screens', *Phys. Med. Biol.* 40, 877.
[20] Eun Hyuk Shin, Youngyih Han, Sang Gyu Ju, Tae Ho Woo, Seung Jae Huh, 'Feasibility study on the real-time motion compensating radiation therapy' the 4th Shinji Takahashi Memorial International Workshop on

3 Dimensional Conformal Radiotherapy (3D-CRT), Nagoya, Japan, on December 10 - 12, 2004.

[21] Youngyih Han, Sang Gyu Ju, Eun Hyuk Shin, Tae Ho Woo, Seung Jae Huh, 'Feasibility Study on the Real-time Motion Compensating Radiation Therapy' The Journal of the Korean Society for Therapeutic Radiation and Oncology', Vol. 22, Suppl. 2, Oct. 2004.

[22] M. S. Muthuswamy, K. L. Lam, 1999. A Method of Beam-Couch Intersection Detection, *Med. Phys.*, 26 (2).

[23] C. N. Coleman, 2002. Radiation Oncology-Linking Technology and Biology in the Treatment of Cancer, Taylor and Francis.

[24] S. Shimizu, H. Shirato, 2001. Detection of Lung Tumor Movement in Real-Time Tumor-Tracking Radiotherapy, *Int'l J. Rad. Onc. Biol. Phys*, 1; 51 (2):304-10.

[25] H. D. Kubo, L. Wang, 2002. Introduction of Audio Gating to Further Reduce Organ Motion in Breathing Synchronized Radiotherapy, *Med. Phys.*, 29(3).

[26] J. T. Booth, S. F. Zavgorodni, 2001. Modeling the dosimetric Consequences of Organ Motion at CT Imaging on Radiotherapy Treatment Planning, *Phys. Med. Biol.*, 46(5):1369-77.

[27] The Development of Quality Assurance Standards for the Remote Controlled Afterloading Brachytherapy Systems and Radiation Source, Korean Food and Drug Administration (KFDA), Nov. 2004.

[28] American Association of Physicists in Medicine, 1997. 'Code of practice for brachytherapy physics: TG No. 56'.

[29] American Association of Physicists in Medicine, 1994. 'Comprehensive QA for radiation oncology: TG No. 40'.

[30] American Association of Physicists in Medicine, 1995. 'Dosimetry of interstitial brachytherapy sources: TG No. 43'.

[31] Khan, 2003. 'The physics of Radiotherapy' 3rd ed, Lippincott Williams and Wilkens, Philadelpia PA.

[32] IAEA, 2000. 'Lessons learned from accidental exposures in radiotherapy', safety reports series No. 17, IAEA Vienna.

[33] Jacob Van dyk editor, 1999. 'The modern Technology of Radiation Oncology' Medical Physics Publishing, Madison, Wisconsin.

[34] Taeho Woo, Hyosung Cho, 'Absorbed dose using nano-scale ion beam for bragg-peak calibrations', *Int. J. Low Radiation*, Vol. 7, No. 3, 236-244, 2010.

[35] Fuminori Soga, 2001. Progress of Heavy Ion Therapy, http://villaolmo.mib.infn.it/Manuscripts/3_medical_applications/Fuminori_Soga.pdf.
[36] The Heidelberg Ion-Beam Therapy Center (HIT) brochure, 2007.
[37] J. F. Ziegler, J. P. Biersack, U. Little mark, 1985, 2003. The Stopping and Range of Ions in Solids, Pergamon Press, New York.

PROBLEMS

1. Let's consider the Semiconductor Photodiodes. If there is a particular PIN photodiode with a pulse of light containing 5×10^{12} incident photons at wavelength of 1.70 μm which gives rise to, on average, 1.5×10^{12} electrons collected at the terminals of the device.

a) Find the energy incident to the photodiode. In addition, calculate the quantum efficiency of the photodiode.
b) It is defined that the diffusion length of a charge carrier is the distance of travelling before it recombines with another oppositely charged carrier and stops carrying any current. The diffusion length of charge carriers is 0.5 μm in this detector. The detector has a circle with a 0.5 μm radius. If the electron diffusion velocity is 5×10^6 cm/s, find the response time of the detector.
c) If the thickness of the intrinsic layer in the photodiode is about 2.5 μm, and the drift velocity of the electrons in this region is 10^5 cm/s. Calculate the response time of the detector.
d) What is the response time of the detector taking drift and diffusion into account?

Solution

a) It is calculated the energy in each photon is (5×10^{-4}) eV/nm × 1,700 nm = 0.85 eV. In addition, the incident energy of the pulse is 0.85 eV/photon × 5×10^{12} photons = 4.25×10^{12} eV = 2.656×10^{-7} Joules. Therefore, it is just equal to 1.5 / 5 = 0.3.
b) A charge carrier of hole or electron travels 0.5 μm at the diffusion velocity to reach an electrode, and this will take 0.5×10^{-6} m / (5×10^4 m/s) ~ 1.0×10^{-10} s = 10 ps.

c) The electrons drift through 2.5×10^{-6} m at a speed of 10^5 m/s, and the time this would take is 2.5×10^{-11} s = 25 ps.
d) It is shown the carriers drift through the active layer and then diffuse to an electrode once they reach the doped semiconductors. Hence, the response time of the detector is the sum of the drift and diffusion response times, or approximately $10 + 25 = 35$ ps.

2. Using the telescopes with singlet objective lenses, resolutions of 0.2 arc min are obtained for visual observations in white light. The light collection efficiency is determined by the biggest piece of glass available. If the lens diameter is 120 mm, find the focal length to reduce the transverse axial chromatic aberration of this objective to the point where this resolution is obtained using typical values for an ordinary glass type.

3. Using the following equation to find the r in the case 45 of θ in the Table 1.

$$D(r,\theta) = \frac{\Lambda \cdot S_K \cdot g(r) \cdot \phi_{an} \cdot 60}{r^2}$$

Table 1.

Var.	r	θ	$g(r)$	ϕ_{an}	Cal. $D(r, \theta)$	RTP $D(r,\theta)$	Dif. (%)
P0	1.9990	0.0000	1.0105	0.6450	79.2162	92.4	-14.27
P1	2.8277	45.0371	1.0149	0.9607	59.2186	59.3	-0.14
P2	2.8277	44.9629	1.0149	0.9607	59.2185	56.7	4.33
P3	2.8277	45.0372	1.0149	0.9606	59.2185	59.0	0.37
P4	1.9990	0.0000	1.0105	0.6450	79.2162	88.4	-10.39
P5	4.9990	0.0000	0.9956	0.6960	13.4665	14.6	-7.76
P6	4.4721	26.5843	1.0047	0.8947	21.8361	22.1	-1.19
P7	4.4721	26.5843	1.0047	0.8947	21.8361	22.2	-1.64
P8	4.4721	26.5843	1.0047	0.8947	21.8361	22.1	-1.19

4.
(a) If a hockey puck weighs 6 ounces (170 g) and can reach the speeds of 120+ mph (192 km/h), calculate the de Broglie wavelength of this puck. Which length scale would exhibit wave-like properties for this puck?

(b) Think its de Broglie wavelength would have to be less than 10^{-14} m in an electron to be confined to a nucleus.

1) What is the kinetic energy of an electron in this region?
2) Is it possible for an election to be in a nucleus?

(c) What is the potential energy difference for an election to make the de Broglie wavelength of 2.5×10^{-10} m? What is the Bragg peak angle for an electron beam on Cu surface?

Chapter 4

NANO TECHNOLOGY (NT)

4.1. ABSTRACT

The nano-scale technology is applied to nuclear industry in the molecular level. The nano-scale state is the most fundamental property of the material. So, the main idea of the NT is focusing on how to manipulate the material in the level of the molecular level. This can be used for the ion interactions of the material in the nuclear material and nuclear waste form. The nuclear reactor theory and thermal-hydraulics could be applied by this new research aspect. In addition, the lunar nuclear power plant is considered where the gravity is very low and the environment has nano-scale atomic contents. The political consideration is investigated, too.

4.2. GOVERNING EQUATION

There are fundamental equations in the nano-scopic description. The mechanics has the governing equations as the momentum equation, continuity equation, and energy equation. These are used in the thermodynamics science and engineering. The conventional equations in solving the application problems are imported to the nano-scale science and engineering. But, it is an important point to find the applicable characteristics in the real applications. It is also necessary to modify the suitable apparatus in the interested stuff.

In momentum equation and continuity equation, Navier-Stokes equations are below. For the incompressible fluid of Newtonian fluid,

$$\rho\left(\frac{\partial v}{\partial t}+v\bullet\nabla v\right)=-\nabla p+\mu\nabla^{2}v+f \tag{1}$$

For cartesian coordinate, momentum equation is, with u, v, w,

$$\rho\left(\frac{\partial u}{\partial t}+u\frac{\partial u}{\partial x}+v\frac{\partial u}{\partial y}+w\frac{\partial u}{\partial z}\right)=-\frac{\partial p}{\partial x}+\mu\left(\frac{\partial^{2}u}{\partial x^{2}}+v\frac{\partial^{2}u}{\partial y^{2}}+w\frac{\partial^{2}u}{\partial z^{2}}\right)+\rho g_{x}$$

$$\rho\left(\frac{\partial v}{\partial t}+u\frac{\partial v}{\partial x}+v\frac{\partial v}{\partial y}+w\frac{\partial v}{\partial z}\right)=-\frac{\partial p}{\partial y}+\mu\left(\frac{\partial^{2}v}{\partial x^{2}}+v\frac{\partial^{2}v}{\partial y^{2}}+w\frac{\partial^{2}v}{\partial z^{2}}\right)+\rho g_{y}$$

$$\rho\left(\frac{\partial w}{\partial t}+u\frac{\partial w}{\partial x}+v\frac{\partial w}{\partial y}+w\frac{\partial w}{\partial z}\right)=-\frac{\partial p}{\partial z}+\mu\left(\frac{\partial^{2}w}{\partial x^{2}}+v\frac{\partial^{2}w}{\partial y^{2}}+w\frac{\partial^{2}w}{\partial z^{2}}\right)+\rho g_{z} \tag{2}$$

Continuity equation is,

$$\frac{\partial u}{\partial x}+\frac{\partial v}{\partial y}+\frac{\partial w}{\partial z}=0 \tag{3}$$

For cylindrical coordinate, momentum equation is, with r, θ, z,

$$\rho\left(\frac{\partial u_{r}}{\partial t}+u_{r}\frac{\partial u_{r}}{\partial r}+\frac{u_{\theta}}{r}\frac{\partial u_{r}}{\partial\theta}+u_{z}\frac{\partial u_{r}}{\partial z}-\frac{u_{\theta}^{2}}{r}\right)$$
$$=-\frac{\partial p}{\partial r}+\mu\left(\frac{1}{r}\frac{\partial}{\partial r}\left(r\frac{\partial u_{r}}{\partial r}\right)+\frac{1}{r^{2}}\frac{\partial^{2}u_{r}}{\partial\theta^{2}}+\frac{\partial^{2}u_{r}}{\partial z^{2}}-\frac{u_{r}}{r^{2}}-\frac{2}{r^{2}}\frac{\partial u_{\theta}}{\partial\theta}\right)+\rho g_{r}$$

$$\rho\left(\frac{\partial u_{\theta}}{\partial t}+u_{r}\frac{\partial u_{\theta}}{\partial r}+\frac{u_{\theta}}{r}\frac{\partial u_{\theta}}{\partial\theta}+u_{z}\frac{\partial u_{\theta}}{\partial z}+\frac{u_{r}u_{\theta}}{r}\right) \tag{4}$$
$$=-\frac{1}{r}\frac{\partial p}{\partial\theta}+\mu\left(\frac{1}{r}\frac{\partial}{\partial r}\left(r\frac{\partial u_{\theta}}{\partial r}\right)+\frac{1}{r^{2}}\frac{\partial^{2}u_{\theta}}{\partial\theta^{2}}+\frac{\partial^{2}u_{\theta}}{\partial z^{2}}+\frac{2}{r^{2}}\frac{\partial u_{r}}{\partial\theta}-\frac{u_{\theta}}{r^{2}}\right)+\rho g_{\theta}$$

$$\rho\left(\frac{\partial u_{z}}{\partial t}+u_{r}\frac{\partial u_{z}}{\partial r}+\frac{u_{\theta}}{r}\frac{\partial u_{z}}{\partial\theta}+u_{z}\frac{\partial u_{z}}{\partial z}\right)$$
$$=-\frac{\partial p}{\partial z}+\mu\left(\frac{1}{r}\frac{\partial}{\partial r}\left(r\frac{\partial u_{z}}{\partial r}\right)+\frac{1}{r^{2}}\frac{\partial^{2}u_{z}}{\partial\theta^{2}}+\frac{\partial^{2}u_{z}}{\partial z^{2}}\right)+\rho g_{z}$$

Continuity equation is,

$$\frac{1}{r}\frac{\partial}{\partial x}(ru_r) + \frac{1}{r}\frac{\partial u_\theta}{\partial \theta} + \frac{\partial u_z}{\partial z} = 0 \qquad (5)$$

For spherical coordinate, momentum equation is, with r, θ, φ (φ = colatitudes),

$$\rho\left(\frac{\partial u_r}{\partial t} + u_r\frac{\partial u_r}{\partial r} + \frac{u_\theta}{r\sin(\varphi)}\frac{\partial u_r}{\partial \theta} + \frac{u_\varphi}{r}\frac{\partial u_r}{\partial \varphi} - \frac{u_\theta^2 + u_\varphi^2}{r}\right)$$
$$= -\frac{\partial p}{\partial r} + \rho g_r + \mu\left[\frac{1}{r^2}\frac{\partial}{\partial r}\left(r^2\frac{\partial u_r}{\partial r}\right) + \frac{1}{r^2\sin(\varphi)^2}\frac{\partial^2 u_r}{\partial \theta^2}\right.$$
$$\left. + \frac{1}{r^2\sin(\varphi)}\frac{\partial}{\partial \varphi}\left(\sin(\varphi)\frac{\partial u_r}{\partial \varphi}\right) - 2\frac{u_r + \frac{\partial u_\varphi}{\partial \varphi} + u_\varphi\cot(\varphi)}{r^2} - \frac{2}{r^2\sin(\varphi)}\frac{\partial u_\theta}{\partial \theta}\right]$$

$$\rho\left(\frac{\partial u_\theta}{\partial t} + u_r\frac{\partial u_\theta}{\partial r} + \frac{u_\theta}{r\sin(\varphi)}\frac{\partial u_\theta}{\partial \theta} + \frac{u_\varphi}{r}\frac{\partial u_\theta}{\partial \varphi} - \frac{u_r u_\theta + u_\theta u_\varphi\cot(\varphi)}{r}\right)$$
$$= -\frac{1}{r\sin(\varphi)}\frac{\partial p}{\partial \theta} + \rho g_\theta + \mu\left[\frac{1}{r^2}\frac{\partial}{\partial r}\left(r^2\frac{\partial u_\theta}{\partial r}\right)\right.$$
$$\left. + \frac{1}{r^2\sin(\varphi)^2}\frac{\partial^2 u_\theta}{\partial \theta^2} + \frac{1}{r^2\sin(\varphi)}\frac{\partial}{\partial \varphi}\left(\sin(\varphi)\frac{\partial u_\theta}{\partial \varphi}\right) + \frac{2\frac{\partial u_r}{\partial \theta} + 2\cos(\varphi)\frac{\partial u_\varphi}{\partial \theta} + u_\theta}{r^2\sin(\varphi)^2}\right]$$

$$\rho\left(\frac{\partial u_\varphi}{\partial t} + u_r\frac{\partial u_\varphi}{\partial r} + \frac{u_\theta}{r\sin(\varphi)}\frac{\partial u_\varphi}{\partial \theta} + \frac{u_\varphi}{r}\frac{\partial u_\varphi}{\partial \varphi} - \frac{u_r u_\varphi - u_\theta^2\cot(\varphi)}{r}\right) \qquad (6)$$
$$= -\frac{1}{r}\frac{\partial p}{\partial \varphi} + \rho g_\varphi + \mu\left[\frac{1}{r^2}\frac{\partial}{\partial r}\left(r^2\frac{\partial u_\varphi}{\partial r}\right) + \frac{1}{r^2\sin(\varphi)^2}\frac{\partial^2 u_\varphi}{\partial \theta^2}\right.$$
$$\left. + \frac{1}{r^2\sin(\varphi)}\frac{\partial}{\partial \varphi}\left(\sin(\varphi)\frac{\partial u_\varphi}{\partial \varphi}\right) + \frac{2}{r^2}\frac{\partial u_r}{\partial \varphi} - \frac{u_\varphi + 2\cos(\varphi)\frac{\partial u_\theta}{\partial \theta}}{r^2\sin(\varphi)^2}\right]$$

Continuity equation is,

$$\frac{1}{r^2}\frac{\partial}{\partial r}(r^2 u_r) + \frac{1}{r\sin(\varphi)}\frac{\partial u_\theta}{\partial \theta} + \frac{1}{r\sin(\varphi)}\frac{\partial}{\partial \varphi}(r\sin(\varphi)u_\varphi) = 0 \quad (7)$$

The energy equation is as follows;

$$\frac{\partial}{\partial t}(\rho E) + \nabla \bullet (\vec{v}(\rho E + P)) = \nabla \bullet (k_{eff}\nabla T) + S_h \quad (8)$$

4.3. NUCLEAR MATERIAL – LOW ENERGY NUCLEAR REACTION (LENR)

4.3.1. Nuclear Material – Vacuum Assisted Lenr

4.3.1.1. Background

The LENR (*similarly*, room temperature nuclear reaction, catalysis nuclear fusion or Cold Fusion) has been investigated for the vacuum assisted situation. The industrial private company has progressed for the commercialization, although the concept of the reaction is not manifested theoretically [1]. By the worldwide nano-technology promotions, the funds have been invested into this research field. The government also starts to recognize the reality of the new concept of the LENR.

The new kind of cold nuclear fusion system is suggested by the Mitsubishi Heavy Industries, Ltd in Japan [2, 3]. New work shows the reliability of the reaction system using simple computer simulation using SRIM 2010 [4, 5, 6]. The displacement can give the possible place for the reaction of the proton (or deuteron) and the interested substrates. The injection gas is sucked by the vacuum in the other side of the substrates. There are unexplained elements in the other side. These could be the products for nuclear reactions because these are non-chemical reaction outputs. There are many stories about the LENR possibilities. Although most books have showed the positive aspects of the LENR, some books have the different estimations.

4.3.1.2. Method and Result

The ion scale power cell application was suggested by Dr. Iwamura in Mitsubishi heavy industry of Japan. The configuration is in Figure 4.1. The decrease of $^{133}_{55}Cs$ makes the nuclear reaction of the increase of $^{141}_{59}Pr$. D_2 gas is injected by the vacuum suction except high energy plasma. The Figure 4.2 shows the collision events in the molecular level. The maximum value of vacuum is around 20 keV and below 150 nm molecular scale. The CaO doesn't affect to the output of displacement in the palladium layer. The Figure 4.3 shows the vacancy distribution in the layers which is considered in the energy production areas.

The vacuum assisted Low Energy Nuclear Reaction (LENR) has been investigated [1]. This can make the possibility of nuclear reaction in the room temperature condition where the theory is not known yet. The computer simulation can predict the reliability of the industrial application in several outputs like ion distribution, range, ionization, phonons, recoiled energy, and damage event.

The SRIM 2010 (Stopping and Range of Ions in Matter 2010) code shows the above variables for the energy production in thin film layers. More reasonable research has been done for the LENR. The displacement of lay ions could be a place for the LENR and make the energy production. Non-quantum mechanical scenario is assumed.

The layers of the substrates are composed of Cr coated thin film Pd (400 Å), CaO and Pd (1000 Å), and bulked Pd (0.1 *mm*). In the case of deuteron (proton), the gas is injected into the surface of the substrates and the other side is vacuumed. The gas of deuteron (proton) is sucked into the other vacuumed place. So, there are many chances of the interaction between deuteron (proton) and the substrate materials.

According to Iwamura [2, 3], the $^{133}_{55}Cs$ is transformed to $^{141}_{59}Pr$. The mass number and atomic number are increased in the case of deuterium gas. This is a proof of the nuclear reaction. The quantity was analyzed by the *X*-ray photoelectron spectroscopy (XPS) for the decrease of $^{133}_{55}Cs$ and the increase of $^{141}_{59}Pr$.

However, it is not assumed to find out the theoretical explanations of the reaction. One can think it is necessary to make the space for the reaction, which is produced by the non-natural conditions like the vacuum suction in the condition of lower energy than the conventional very hot nuclear reaction. The

CaO is another condition for the room temperature reaction. This could be changed to other material.

The SRIM 2010 code is used [4, 5]. This is a group of programs which can calculate the stopping and range of ions from 10 eV to 2 GeV/amu in the quantum mechanical matters including solid and liquid states for the ion-atom collisions. The atoms and ions are screened by the coulomb collision. This new code is improved comparing to the previous codes; SRIM 2000 and TRIM 96 (Transport of Ions in Matter; this is the former name of the SRIM code system). The improved points are that the minor bugs are removed which was in the initial version of the code and it is well modified for the heavy ions. In addition, the nuclear stopping power for the specific atoms is introduced. In the real situation, the irradiation effects are instantaneous which need more complex methods to develop experiment.

For the simulation, the small energy in injected instead of the suction of the gas in the vacuum. So, the ion beam is used for the simulation. The experimental system is able to be suggested in the different system for the vacuum assisted structures. However, fundamentally vacuum is the main factor in the non-hot nuclear reaction. The Figure 4.1 shows the system of the Iwamura. The Figure 4.2 shows the production of the displacement, which is assumed for the reaction. The Figure 4.3 shows the maximum value of the vacuums around 20 keV. The CaO layer doesn't affect the output of the displacement in the Palladium layer. So, there is not a dependency of the CaO quantities in making the place for the nuclear reaction in the room temperature. There are some results of simulation in Figure 4.3. Table 4.1 shows the evolution of LENR.

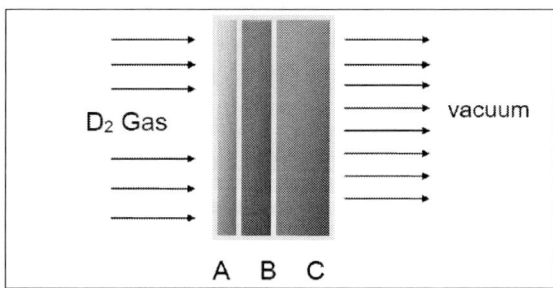

Figure 4.1. The layers of Pd complex system by Iwamura. A: Cs or Sr + Pd(400Å), B:CaO and Pd(1000Å), C:Pd(0.1mm).

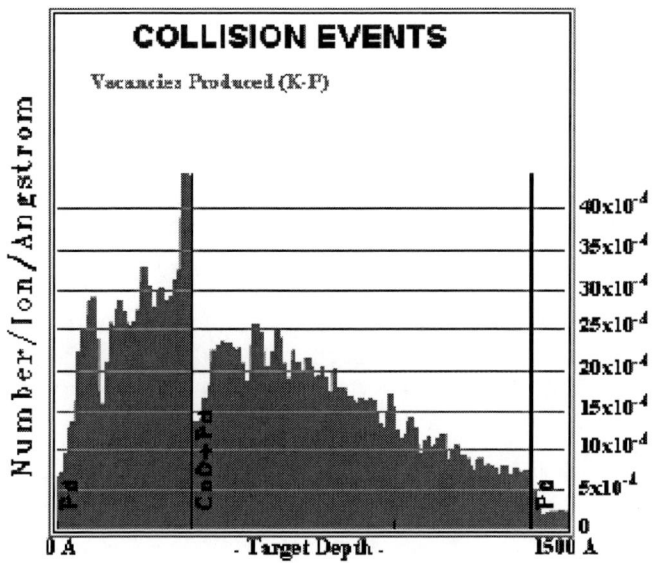

Figure 4.2. The collision event of SRIM 2010.

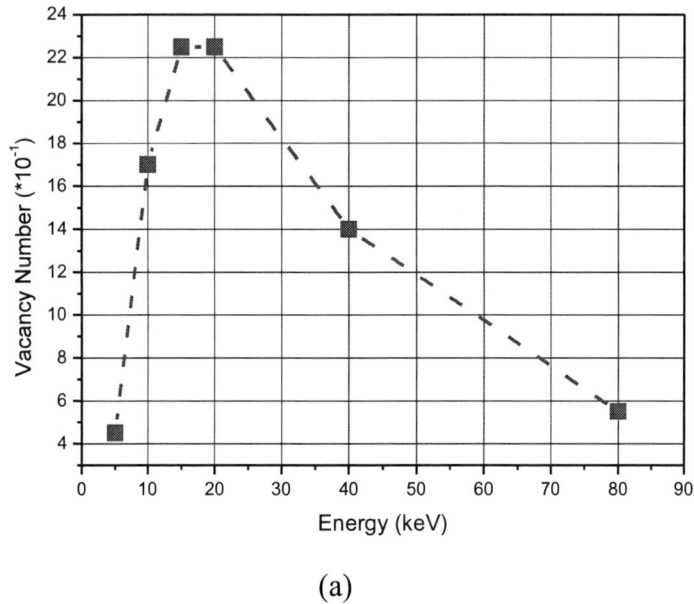

(a)

Figure 4.3. Continued.

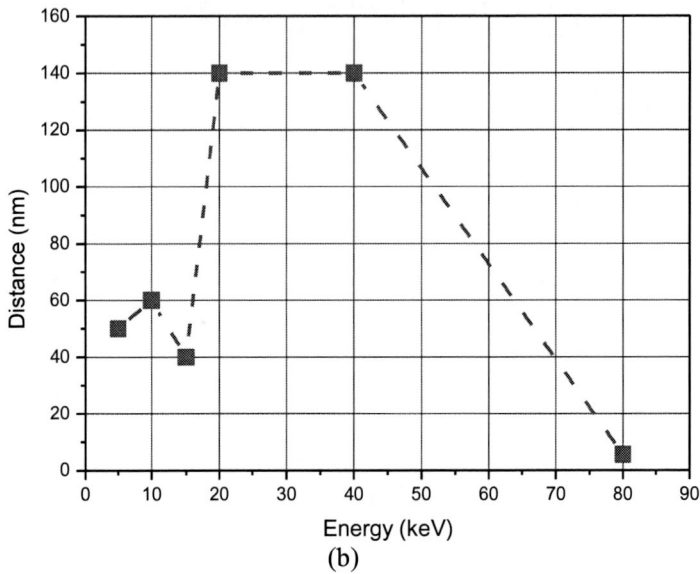

(b)

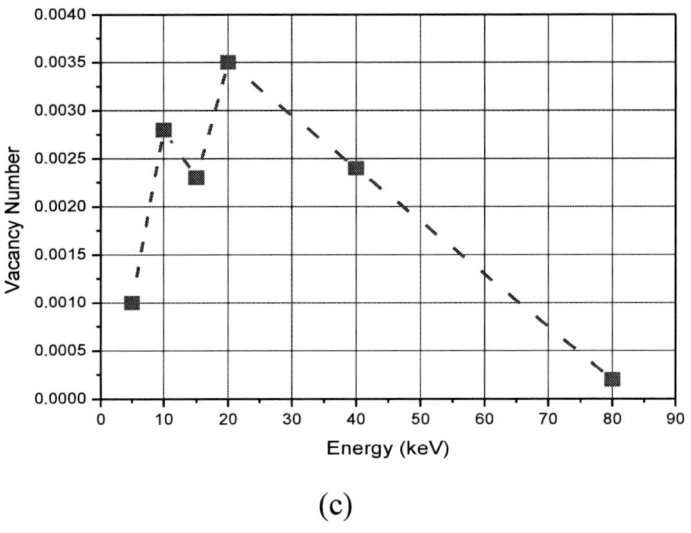

(c)

Figure 4.3. Vacancy per D+ collision.

Table 4.1. The evolution of the LENR

Type	Year	LENR initiator
1st	1989 ~	Electricity energy (Electrolysis)
2nd	2002 ~	Sound wave energy (Sonoluminescence)
3rd	2003 ~	Non-energy physical properties (Vacuum, etc.)

4.3.1.3 Summary

In the Vacuum assisted LENR, the complex is an important factor. Namely, such as the palladium-platinum is the best couple for the reaction in LENR, CaO layer should be in sandwich position between the palladium layers. However, it is a main point to reduce the vacuum energy. This is the meaning of the reducing the work energy for the room temperature reaction.

4.4. NUCLEAR WASTE MANAGEMENT

4.4.1. Background

The nuclear waste business has an important turning point in the country during the new repository selection [7]. It is necessary to develop the more reasonable and safe methods for the nuclear waste confinement. The micro and nano-scale investigation is a new frontier for the object of the research. The advanced small scale technology should be developed for this waste issue. In Figure 4.4, the ion beam distribution is shown. The highest point for the critical dose is at 300 nm. Crystalline Silicotitanate is high for stopping power of most regions among 4 forms in Figure 4.5. The simulation is done that 10,000 ions are injected into the 4 targets. This makes the displacement in the target material. The comments for the conclusion of this simulation are given. The susceptibility is found as the characteristics of the waste forms. The experimental research should be accompanied with this theoretical simulation research. The critical dose and energy loss are examined for the suitable waste forms.

The irradiation induced nuclear waste material is investigated. The material science and geological aspects are considered for the 4 materials which are used for the nuclear waste forms. The alpha particle is considered as the irradiated radiation. Four materials are susceptible to the irradiation-induced amorphization. Several variables are investigated for the ion-radiation interactions. The Stopping and Range of Ion in Matter 2010 (SRIM 2010) code system is used to show that the ion dose is changed to the displacement per atom (dpa) completely and the kinetic energy is transferred to each target atom through nuclear collision. The necessary thickness of the waste form is investigated. More reasonable research has been done for the nuclear waste material container.

4.4.2. Method

The geological investigation of nuclear wastes is studied for the more real descriptions in the radioactive waste management. The nuclear repository has been confronting for the unreasonable problems. One of these matters is the very long and unimaginable reliability guarantee of the nuclear waste repository. It is necessary to show at least 1,000 years safety confirmation of the site. Conventional nuclear waste research has been focusing on the simulations of the releases of the unwanted scenarios following the assumed variables like the geometry of waste source, the groundwater, and so on. However, this kind of research couldn't give the explanation of the real situations which might be happened in the microscopic or nano-scopic scale fields. Unless the full research in the very small world is done, the nuclear release oriented diffusion aspects research is not reasonable. Especially, the Transuranic (TRU) waste is studied for the alpha-emitting radionuclides of the sufficiently long life of atomic number 92 and concentrations greater than 100 nanocuries except ^{289}Pu and ^{241}Pu and their daughter products (sometimes these are considered as TRU waste by local requirement.)[8]. Fortunately, we are in the beneficial atmosphere of the nano-scale research in the new century. This can realize the long term safety guarantee, because the nano-scale world is monitored by the new technology. This can protect the people from the macroscopic nuclear materials diffusion accidents in the long term. This research can be done by the HRTEM (High Resolution Transmission Electron Microscopy), SEM (Scanning Electron Microscope), and some other solid state physical measurement instrumentation. In addition, the remarkable development of computation tool increases the reliability of the estimation of the nano-scopic or microscopic behavior of the nuclear wastes. This is done by similar ways of molecular simulation in the multi-scale aspects. Therefore, if all of these investigations have accomplished, the waste site selection problem could be solved by the residents' understanding this new reliable research trend.

The word 'metamict' or 'metamictization' is used as synonymous for 'amorphous' or 'amorphization' in the heavy particle irradiation like alpha particle event damage in the mineral and ceramics. The wastes forms are fixed in a glass matrix or the synthetic rock form [9].

The SRIM 2010 code is used for the simulations [10, 11]. This is a group of programs which can calculate the stopping and range of ions from 10 eV to 2 GeV/amu in the quantum mechanical matters including solid and liquid states for the ion-atom collisions. The atoms and ions are screened by the coulomb collision. This includes the exchange and correlation interactions between the overlapping electron shells. The ions have the long interactions with target atoms which create the electron exciting and plasmons within target. These are described by including a description of the target's collective electron structure and inter-atomic bond structure when the calculation is setup. The charge of the ions within target is expressed by the concept of effective charge, which includes a velocity dependent charge state and long range screening because of the collective electron sea of the target. In the real situation, the irradiation effects are instantaneous which need more complex methods to develop experiment [6, 7].

4.4.3. Result

In this chapter, 4 kinds of materials were investigated; Sheet Silicate (Mica) compositions, Zeolite compositions, Smectite compositions, and Crystalline Silicotitanate. Table 4.2 shows the examples of the materials and Table 4.3 show the critical doses. The Nontronite of Smectite has the highest value. The highest deposition is around 300 nm. The alpha particle is injected ion in the simulation [12]. For the susceptibility, it is suggested to calculate following equation [13, 14]. The number of displacements per atom of the target is,

$$\text{dpa} = (1 \times 10^{-16}) N_d \cdot D_c / \rho_n \tag{11}$$

where,

N_d : Target displacement(displacement/ion/Å)
D_c : Critical dose(ions/cm^2)
ρ_n : Atomic density of the target material(atoms/Å^3)

Table 4.2. Waste Form Compositions

Name	Composition
Sheet Silicate (Mica)	
Muscovite	$KAl_2(AlSi_3O_{10})(OH,F)_2$
Phloopite	$KMg_3(AlSi_3O10)(OH,F)_2$
Biotite	$K(Mg,Fe)_3(AlSi_3O_{10})(OH,F)_2$
Lepidolite	$K(Li,Al)_3(Al,Si)_4O_{10}(OH,F)_2$
Zeolite	
Analcime	$Na1_6Al_{16}\,Si_{32}O_{96}\cdot16H_2O$
Stilbite	$NaCa_{16}A_{15}\,Si_{13}O_{36}\cdot16H_2O$
Natrolite	$Na_2Al_2\,Si_3O_{10}\cdot2H_2O$
Zeolite Y	$NaAlSi_2O_6\cdot xH_2O$
Smectite	
Montmorillonite	$Kx(Al_2\text{-}yMgy)Si_4O_{10}(OH)_2\cdot nH_2O$
Nontronite	$Na_{0.3}Fe_2(Si,\,Al)_4O_{10}(OH)_2\cdot nH_2O$
Saponite	$Ca_{0.25}(Mg,Fe)_3(Si,Al)_4O_{10}(OH)_2\cdot nH_2O$
Crystalline Silicotitanate	
Crystalline Silicotitanate	$Na_2Ti_2O_3SiO_4\cdot2H_2O$

Table 4.3. Critical Dose

Waste Form	Critical dose (ions/cm^2) : D_c
Sheet Silicate(Mica) : Muscovite	18.4438×10^{-4}
Zeolite : Analcime	146.9250×10^{-4}
Smectite : Nontronite	217.5820×10^{-4}
Crystalline Silicotitanate	28.8761×10^{-4}

This is used for the calculation for D_c (critical dose, ions/cm^2). This is used for finding the susceptibility of the waste form.

In the Figure 4.4, the ion beam distribution is shown. The highest point for the critical dose is at 300 nm. Crystalline Silicotitanate is highest for stopping power of most regions among 4 forms in Figure 4.5. The simulation is done 10,000 ions are injected into the 4 targets. This makes the displacement in the target material. The comments for the conclusion of this simulation are as follows;

1) The susceptibility is found as the characteristics of the waste forms.
2) The experimental research should be accompanied with this theoretical simulation research.
3) The critical dose and energy loss are examined for the suitable waste forms.

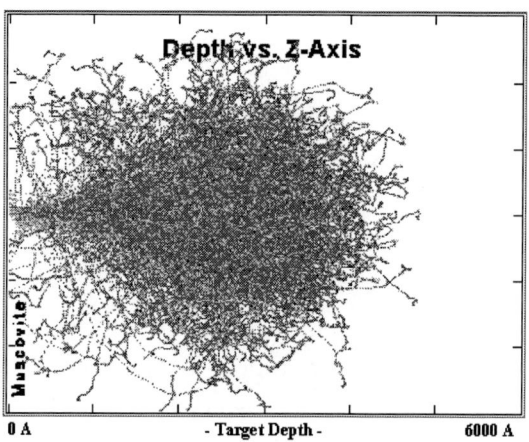

Figure 4.4. Z-axis view of He+2 ion injection for Muscovite.

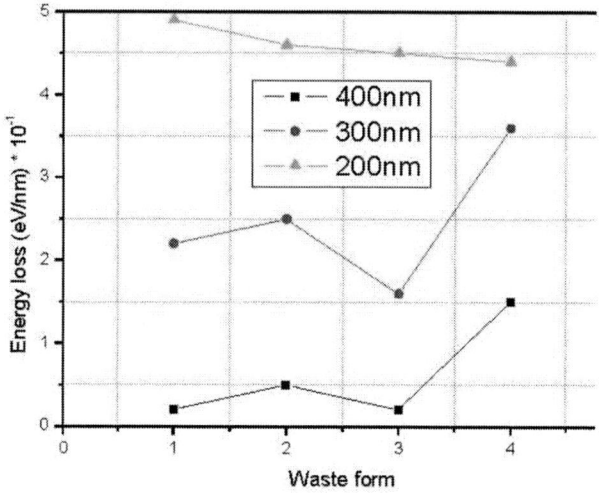

Figure 4.5. Stopping Power (1: Muscovite, 2: Analcim, 3: Nontronite, 4: Crystalline Silicotitanate).

4.4.4. Summary

The nuclear waste business is a good turning point in the country due to the new repository selection. It is necessary to develop the more reasonable and safe methods for the nuclear waste confinement. The micro and nano-scale investigation is a new frontier for the object of the research. The advanced small scale technology should be developed for this waste issue. Table 4.2 is the waste forms. There is the critical dose in Table 4.3.

4.5. LUNAR NUCLEAR POWER PLANT

4.5.1. Background

In the moon, the atmosphere is nearly not existed due to the very low gravity [15]. So, the contents of the environment could be recognized as a molecular nano-scale state and the radiation heat transfer is the major cooling method [16]. The coolant in the nuclear power plant of the lunar base is the moon surface soils [17], which is called the regolith. The regolith means the layer of loose and heterogeneous material covering the solid rock. The optimized length of the radiator for the coolant in the lunar nuclear power plant is related to the produced power and the moon environment temperature. In addition, the length will increase in the case of severe accident where the nuclear fuel and core will be failed. The heat removal will be done without structure failure in severe accident, if the piping of coolant is not failed. This situation is different from the nuclear power plant system in the earth, because heat is removed to environment without any restriction comparing to the condition of earth. Therefore, the plant will be used again without permanent shut down exchanging some component like the damaged core and fuel rods. In this chapter, the safety assessment is performed in order to probabilistic analysis of the severe accident.

Table 4.4 shows the contents of the lunar highland soils with the earth contents [18]. In the moon soil, the contents of aluminum, calcium, magnesium, and chromium are higher than those of the earth. This is the particular characteristic factor of the heat transfer in the moon. The energy is transferred only by the radiation due to the very weak atmosphere. It is very cold in the night and very hot in the daytime on the surface of the ground [19].

Historically, Konstantin Tsiolkovsky made a suggestion for the colony on the moon [20]. A number of ideas for the conceptual design have been proposed by the scientists. Among them, Arthur C. Clarke mentioned a lunar base of inflatable modules covered in lunar dust for insulation in 1954. For the better safety design, John S. Rinehart suggested the structure of the stationary ocean of dust, because there could be a mile-deep dust ocean on the moon [21]. In 1959, the project horizon was launched regarding the U.S.

Army had a plan to establish a fort on the Moon by 1967 [22]. The project was leaded by H. H. Koelle, a German rocket engineer of the Army Ballistic Missile Agency (ABMA).

Table 4.4. Comparison between lunar high land soil and earth averages in ppm

Element	Lunar highland	Earth
Oxygen	446,000	466,000
Silicon	210,000	277,000
Aluminum	133,000	81,300
Iron	48,700	50,000
Calcium	106,800	36,300
Sodium	3,100	28,300
Potassium	800	25,900
Magnesium	45,500	20,900
Titanium	3,100	4,400
Hydrogen	56	1,400
Phosphorous	500	1,050
Manganese	675	950
Carbon	100	200
Chlorine	17	130
Chromium	850	100

The first landing was done in 1965 and 245 tons of cargos were transported to the outpost by 1966. Recently, South Korean government launched the NARO-1 rocket for the space development at Naro Space Center which is located about 485 km from south of Seoul [23]. The current aims of the R&D are the satellite and space launch technology constructions in the space commercial usage. However, following the trend of the global space development challenge, it is proposed to make the theoretical research in the lunar base erection.

4.5.2. Thermal Hydraulic Design

The Figure 4.6 shows the loop of the nuclear power plant with several values. It is assumed as the normal nuclear power plant as the very high temperature reactor (VHTR) where 600 MW$_{th}$ power including 1,300 K as maximum outlet temperature. It is reasonable the gas cooled reactor is possible in the moon comparing to the water cooled reactor because there is no water resource. The heat exchanger is analyzed for the heat transfer. The Figure 4.7 shows the nodalization of the heat exchanger. It is numbered from #1 to #9.

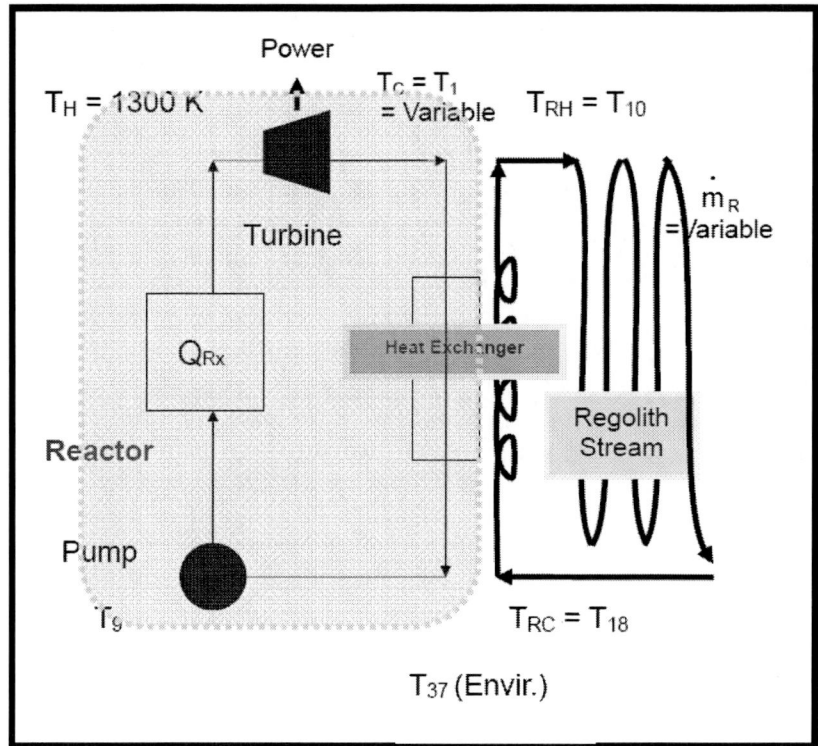

Figure 4.6. Loop of lunar nuclear power plant.

The regolith flow needs the constant environmental temperature in the moon due to the big temperature difference between the daytime and the nighttime. In this paper, the environmental temperature is assumed as the maximum daytime temperature of 396.15 K. The Figure 4.8 shows the regolith flow. Figure 4.9 shows the nodalization of the regolith flow.

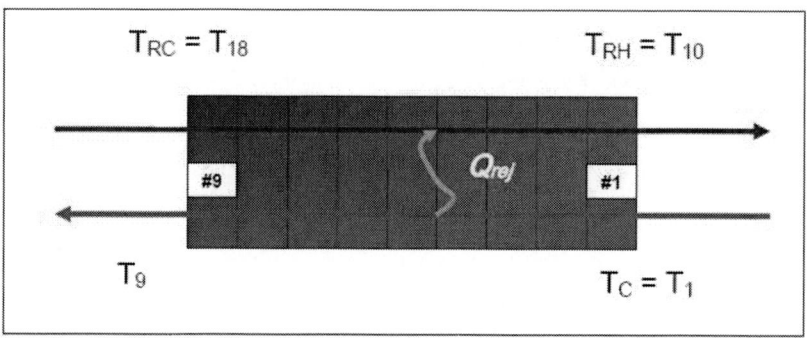

Figure 4.7. Heat exchanger nodalization.

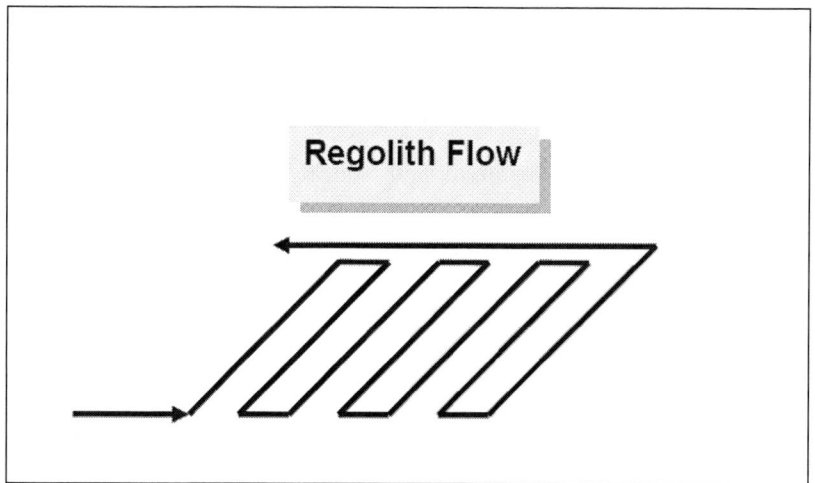

Figure 4.8 Regolith flow.

There are 20 nodals in this chapter. The ambient lunar environment cools the thermal energy from the heat exchanger. The rejection thermal energy from nuclear power reactor is obtained in equation (1).

T_C is changed by the cooling of the heat exchanger which is cooled subsequently by the regolith flow. The heat exchanger has 9 nodals in this model. The energy of the heat exchanging part is transferred by radiation energy and it is assumed as $\varepsilon^{HX} = 1.0$.

$$Q_{rej} = \left(1 - 0.6\left(\frac{T_H - T_C}{T_H}\right)\right)Q_{Rx} = \left(1 - 0.6\left(\frac{1300 - T_C}{1300}\right)\right)Q_{Rx} \quad (1)$$

$$Q_{nodal\#1}^{HX} = \sigma\varepsilon^{HX} A\left(T_1^4 - T_{10}^4\right) \quad (2)$$

$$Q_{rej} = Q_1^{HX} + Q_2^{HX} + Q_3^{HX} + Q_4^{HX} + Q_5^{HX} \\ + Q_6^{HX} + Q_7^{HX} + Q_8^{HX} + Q_9^{HX} \quad (3)$$

The length of the heat exchanger is found by the NTU (Number of Transfer Units). For this chapter, it is modified for the radiation heat transfer case of equation (4) and (5). In the heat exchanger, mass flow rate ($\dot{m}$) is 15.1 kg/s and $C_p^{potassium}$ is 0.8 $kJ/(kg \cdot K)$ for potassium coolant flow. T_1 and T_9 should be same. The temperature and entropy relationship can find the temperature. In Figure 4.10, T_1 (also T_9) can be changed by the thermal efficiency changing.

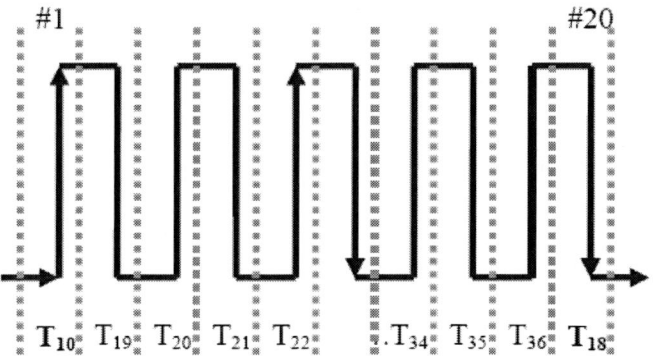

Figure 4.9. Regolith flow nodalization.

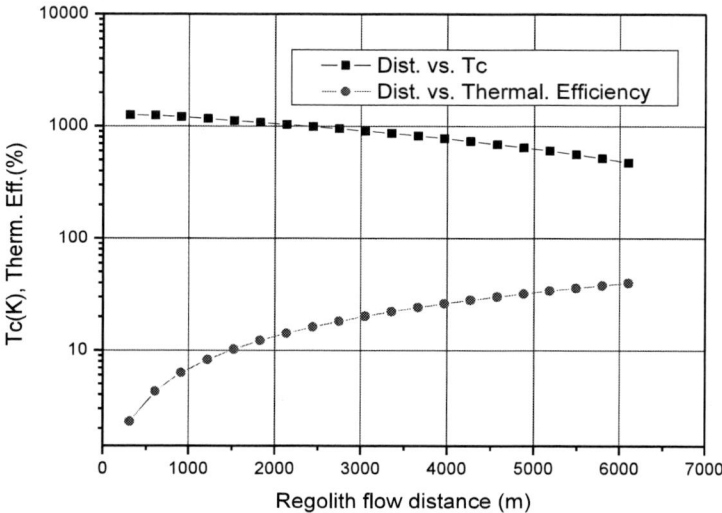

Figure 4.10. Thermal efficiency.

$$Modified\ NTU = \frac{T_1 - T_{18}}{T_9^4 - T_{10}^4} = \frac{1260 - T_{RC}}{1260^4 - T_{RH}^4} \quad (4)$$

$$HXLenght = \frac{\dot{m} C_p^{potassium}}{\sigma \varepsilon^{HX} P} \times Modified\ NTU \quad (5)$$

The radiation energy from the regolith to the lunar environment is described by the hear flux. $Q_{nodal\#1}^{reg}$ is variable by the distance of the regolith flow. There are 20 nodals in this model. The regolith flow of the radiator is;

$$Q_1^{reg} = \dot{m} C_p \left(T_{10} - T_{19} \right) \quad (6)$$

Mass flow rate ($\dot{m}$) is obtained by ρVA of the regolith flow. Density is 0.3 kg/m³. Velocity is variable. Area is calculated by the interaction with the heat conduction equation (6) and the radiation heat transfer equation (7). The area is also iterated with the optimized regolith flow length. For the radiation heat transfer to environment, the nodal # has the rejected heat which is different from the rejected heat of the nodal #20. T_{37} is 396.15 K. Equation (6) shows the rejected heat in each nodal. Finally the T_{10} is obtained in equation (9).

$$Q_1^{reg} = \sigma \varepsilon A \left(T_{10}^4 - T_{37}^4 \right) = \sigma \varepsilon A \left(T_{10}^4 - (396.15)^4 \right) \tag{7}$$

$$Q_{rej} = Q_1^{reg} + Q_2^{reg} + Q_3^{reg} + Q_4^{reg} + Q_5^{reg}$$
$$+ \ldots + Q_{17}^{reg} + Q_{18}^{reg} + Q_{19}^{reg} + Q_{20}^{reg} \tag{8}$$

$$T_{10} = \sqrt[4]{\frac{Q_1^{reg}}{\sigma \varepsilon A} + (396.15)^4} \tag{9}$$

4.5.3. Safety Assessment

Although the new kind of material is suggested as the coolant, it is very unsure to get the reliability of the performance. The conventional method for the success probability is introduced for the lunar base plant. This is the basic scenario of the accident as the failure of the cooling of the modeling. The basic data is used by the failure frequency data of the modified SECY-93-092 [24] in Table 4.5. Figure 4.11 shows the event tree of the failure to the normal cooling. The initiating event is the failure of the cooling path which means any kinds of the cooling failure like the rupture or the blocking in the coolant loop. The event tree shows the consistent reactions against the initiating event. Core trip is the first reaction for the safety consideration. Then, several safety actions and the power equilibrium are responded. Finally, the long-term cooling by the convection is performed. Table 4.6 is the result of the propagation of the safety assessment.

Table 4.5. Modified event likelihood of occurrence based on SECY-93-092

Event	Frequency of Occurrence
Likely events	$> 10^{-2}$ / plant-year
Non-likely events	$10^{-2} \sim 10^{-4}$ / plant-year
Extremely non-likely events	$10^{-4} \sim 10^{-6}$ / plant-year
Very rare events	$< 10^{-6}$ / plant-year

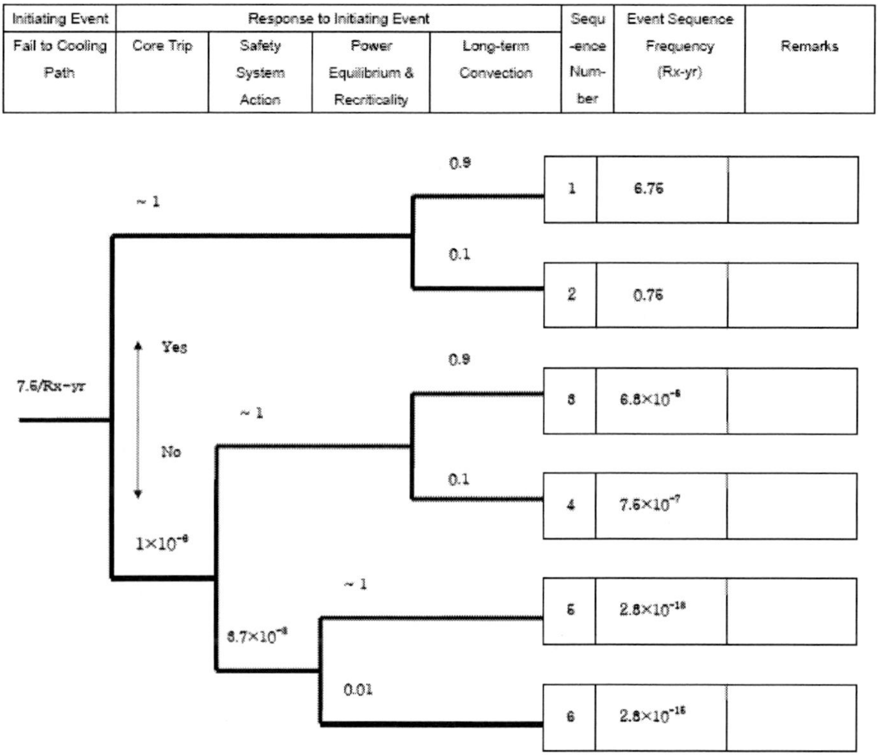
Initiating Event
Response to Initiating Event
Fail to Cooling Path
Core Trip
Safety System Action
Power Equilibrium & Recriticality
Long-term Convection
Sequ-ence Num-ber
Event Sequence Frequency (Rx-yr)
Remarks
7.6/Rx-yr
~ 1
Yes
No
1×10-8
0.9
0.1
~ 1
0.9
0.1
~ 1
0.01
1
6.76
2
0.76
3
4
7.6×10-7
5
6

Figure 4.11. Event Tree of Fail to Normal Cooling.

4.5.4. Result

The regolith of the moon is the optimized coolant matter. This chapter shows the engineering aspect of the thermal facility in the moon where the different kinds of the coolant for the heat exchanger are imagined for the possible nuclear power plant. The state of the nano-scale content atmosphere is nearly regarded as the vacuum state, although there is a factor of weak gravity which is related to the contents of the atmosphere. This means the heat transfer depends on the condition of the gravity in the candidate planet for the human habitat base, because the content of environment is a key factor for the heat transfer.

Table 4.6. Event frequency for cooling (Rx-yr)

Event	Event frequency
1	6.75×10^{0}
2	0.75×10^{0}
3	6.80×10^{-6}
4	7.50×10^{-7}
5	2.80×10^{-18}
6	2.80×10^{-16}

Using the simulation of the above modeling, the results are obtained as follows;

- The optimized length of the condenser is obtained by the rejected energy from the reactor and the radiation thermal energy to the lunar environment. The Figure 4.12 (a) shows that the minimum length of the regolith stream is 6,100 m at the 0.204 m of diameter in the cylindrical shape condenser. The diameter is decided following the length of the regolith flow. Namely, if the length is shorter, the diameter is larger. So, the optimized diameter can be obtained due to the environmental situation. If there is a small space for the regolith cooling pond, the larger diameter is needed.
- The important view of lunar nuclear power is how to control the severe accident which is an accident of core damage like nuclear fuel melt and core failure accident. If this case happens in the Moon, heat will be radiated to space. However, one needs to find the possibility of heat removal without structure failure. So, it is needed to keep the stabilized system with heat rejection. The regolith flow distance is calculated in Figure 4.12 (b), where the length is 32,274 m at the 0.204 m of diameter in the cylindrical shape condenser.

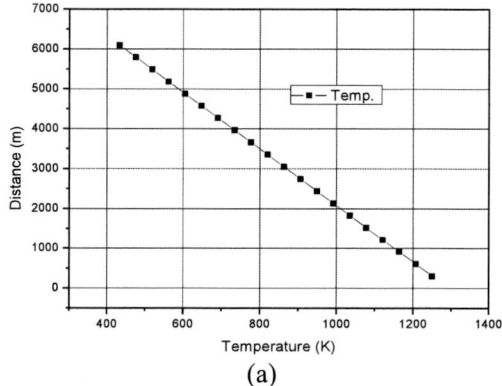

(a)

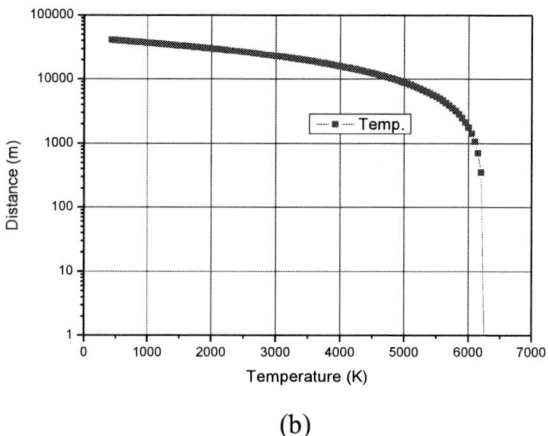

(b)

Figure 4.12. Optimized energy rejection by radiation in regolith flow.(a) normal operation, (b) severe accident case.

- The temperature is assumed as 6,250 K. Usually, fuel melting temperature is assumed more than 5,000 K for the conservative calculation. The melting point of uranium is 1,405.35 K and boiling point is 4,404.15 K. Therefore, the 6,250 K is decided considering an assumption of the conservative value to see the ability of the cooling for the decay heat removal.
- The mass flow rate is changed by the nodal number in Figure 4.13.

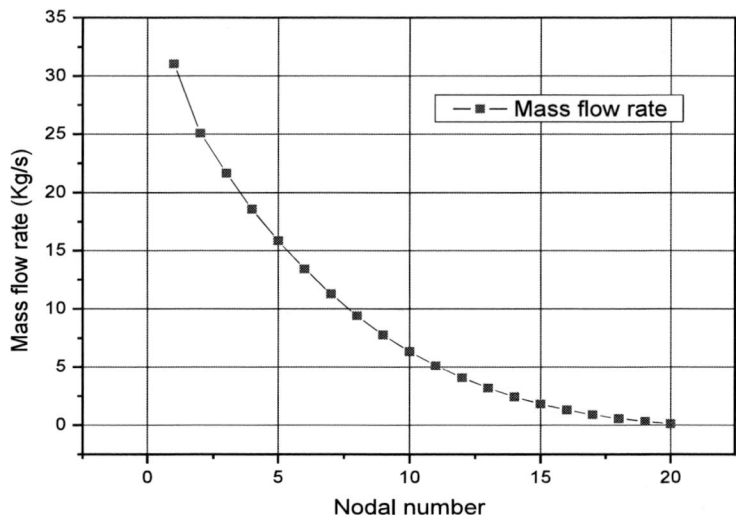

Figure 4.13. Mass flow rate by the nodal number in regolith flow.

- The Figure 4.14 shows the heat exchanger nodalization. The temperature is calculated as the 9 nodals by radiation heat transfer from the potassium coolant to the regolith flow.

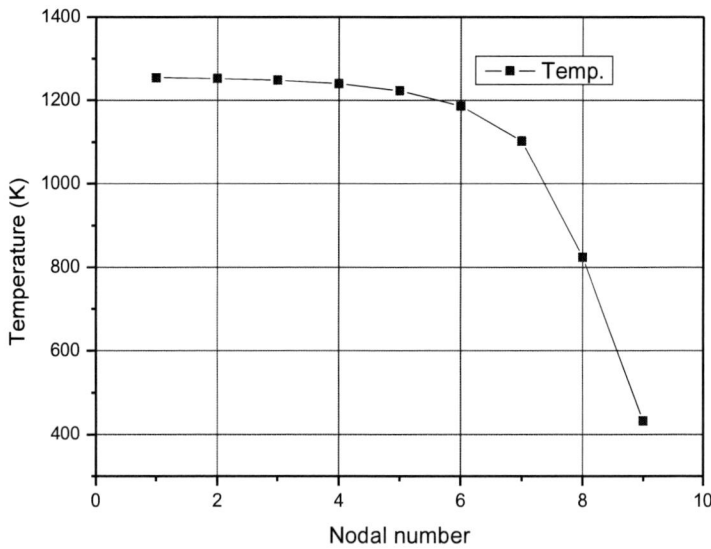

Figure 4.14. Heat exchanger nodalization.

- The thermal efficiency and T_C are changed by the regolith flow length.
- The failure frequency shows that the accident event of the failure to normal cooling is responded as the successful case #1 as 6.75 / Rx-yr against 7.05 / Rx-yr of the initiating event.

Using the safety assessment, the event sequence frequency of the case *#1* is the highest value. The proportional rate between the failure frequency of final event and the failure frequency of initial event is obtained, which is 90.0 % (={6.75 / Rx-yr} / {7.50 / Rx-yr}). This means that the event of the long-term convection happens as 90.0 % probability when the cooling path fails. The other cases are negligible comparing to the case *#1*.

The coolant feature is very important in the construction of the lunar nuclear power plant due to the safety as well as the economical factors. Therefore, the optimized length of the coolant will be one of the critical issues in the operations. The conclusion can be obtained as follows;

- The optimized coolant loop in the moon is constructed.
- The economical heat transfer simulation is performed.
- Hypothetical accident in lunar nuclear power plant is analyzed.
- The safety concern can be proposed in the possible lunar base power plant structure.

4.5.5. Summary

The reasonable length of the lunar nuclear power plant is obtained. Most interested consideration for severe accident is studied. If the structure is stable, the accident of core failure by fuel melting is sufficiently treated by increasing of the regolith flow pipe. For further research, several kinds of power cases could be modeled. The optimized power plant is supposed for the lunar base. The safety assessment for the possible accident scenario is performed for the better reliability of the thermal hydraulic design. The recent news, however, says a lot of water exists in the surfaces on the moon. If the quantity is abundant, the water coolant could be used. There are still the significant considerations of the behaviors of the coolant due to the weak gravity which produces the nano-scale contents of the atmosphere. In addition, the shielding against meteors should be considered in the aspect of engineering standard which could be possible by the strong structures like the concrete steel type building. This technology has been development in the containment of conventional nuclear power plant, which can withstand the crash of airplane or the shaking of earthquake.

4.6. NUCLEAR POLICY – ATOMIC NANOMICS INITIATIVES (ANI)

4.6.1. Background

A new research trend in nuclear science and engineering is investigated [25]. The nano-scale technology has been started for the revolutionary research initiative in the wide range of our lives. The applications in the nuclear industry using nanotechnology are discussed. This promises the new marketing creation with the big profits. For this research business, new concepts and preparations are necessary. The Atomic Nanomics Initiative (ANI) is suggested for the applications of nano-scale matters in nuclear technology. The TechSim 2003 is introduced for the science and technology simulator which is used for the estimation of the future ANI procedures using a Systems Thinking method, System Dynamics (SD), until year 2020. The future performance of this initiative is expected.

4.6.2. Method and Result

The new technology field is addressed using the nanotechnology. The National Nanotechnology Initiative (NNI) has expected to be applied to the wide ranges of science and technology following the technological, political, and economical interests which are invested to this new field. The 21st century is fundamentally different from the previous century in the science and technology which are characterized as the multi-disciplinary oriented industrial innovation. The nanotechnology application in nuclear engineering was introduced in 2002 for the first time in the history of Korean Nuclear Society (KNS) [26]. The Systems Thinking algorithm can suggest the reliable way for the technology procedure of specified areas. This shows in the Figure 4.15. The SD is a well known tool for the completing for the Systems

Thinking process. The simulator TechSim 2003 is expected for the seed concept of other kinds of the technology decision-making algorithms.

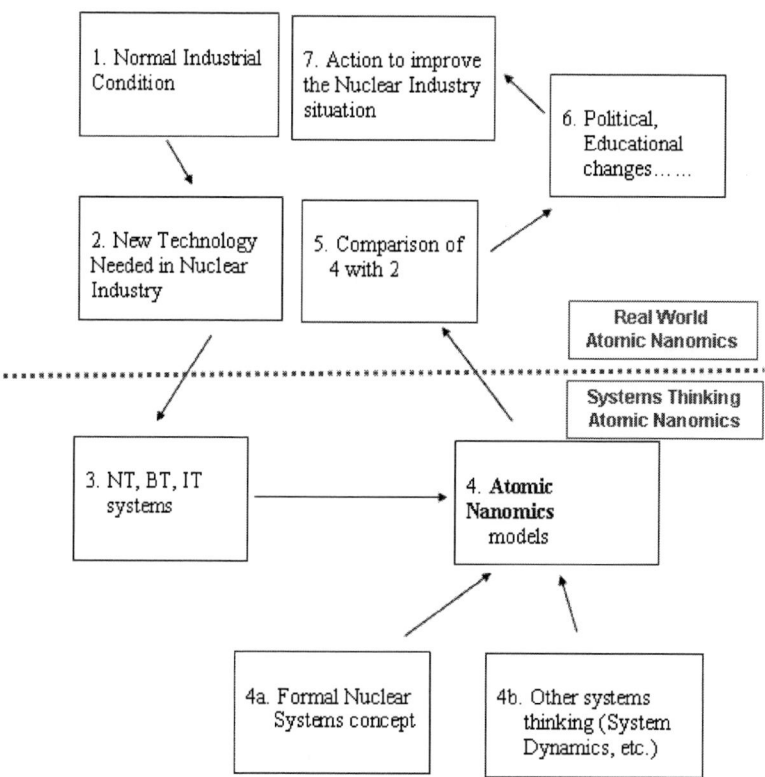

Figure 4.15. The Real World vs. Systems Thinking Atomic Nanomics.

4.6.2.1. *National Nanotechnology Initiative (NNI)*

Albert Einstein, as part of his doctoral dissertation, calculated the size of a single sugar molecular from experimental data on the diffusion of sugar in water. His work showed that each molecule measures about a nanometer in diameter [27]. Later in 1959, Richard Feynman announced, "There is Plenty of Rooms at the Bottom." The nano-scale research was initiated around November 1996 nationally by several governmental members in the United States. The NNI plan by National Science and Technology Council (NSTC) went to President Clinton, and, subsequently, the budget was submitted to congress in 2001 which is referred as the NNI officially [28].

4.6.2.2. Atomic Nanomics Initiative (ANI) System Analysis

The systemic analysis of the ANI is estimated by the TechSim2003. This has been constructed by SD model which is the time step affected Systems Thinking algorithm for the complex systems in the natural problems as well as the social matters [29, 30].

The configuration of the algorithm is shown in the Figure 4.16. The time rages are from 2002 to 2020. The 2020 is the final planning year used in the NNI. The graph is simply connected to three factors as the Technological Factor, Economic Factor, and Political Factor. Each factor is quantified as the random sampling for the positive and negative values. The result is in the Figure 4.17. There are 3 different graphs, where the unit of y-axis is Korean won ($1 = 1,100 won). The NanoFab and the PresidChange are considered factors. The Figure 4.18 shows the sensitivity of the case 'NanoFab + PresidChange'. The 50 %, 75 %, 95 %, and 100 % are classified as it seen by the colors.

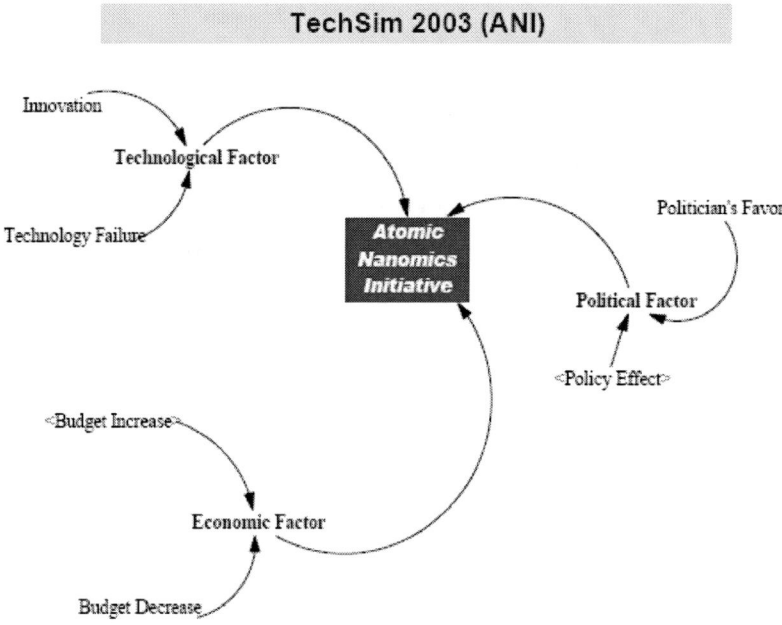

Figure 4.16. The TechSim 2003 for Atomic Nanomics Initiative (ANI).

Nano Technology (NT)

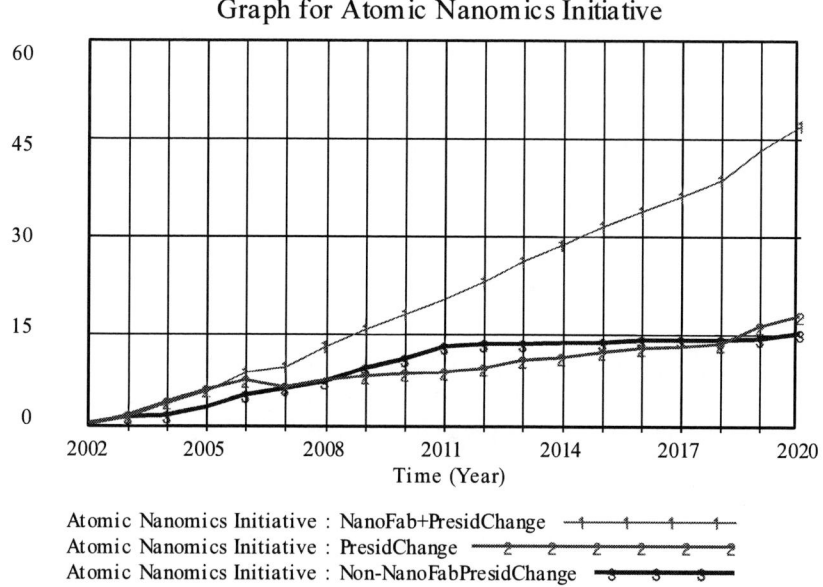

Figure 4.17. The Development Quantification.

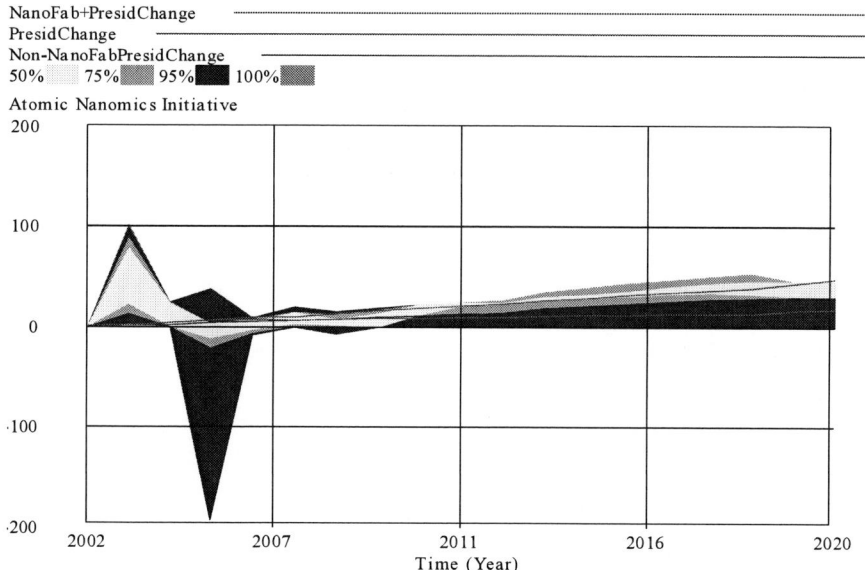

Figure 4.18. The Sensitivity of Case 'NanoFab + PresidChange'.

Therefore, three different technical algorithms are introduced as following descriptions. If there are more factor are considered, it can be constructed similar to one of these methods.

- The Technological Factor : The Integration of randomly sampled two factors(Innovation and Technology Failure)' frequencies is done. - Time independent algorithm.
- The Economic Factor : Budget Increase is integrated by the randomly sampled frequency of NanoFab Opening after year 2005. - Time progress algorithm.
- The Political Factor : Policy Effect's randomly frequency are high when the Presidential election happens in every 5 years as 2002, 2007, 2012, and 2017. - Time periodic algorithm.

The ANI is not much different from the NNI. All concepts are related each other. The ANI is just focused to the nuclear industry. The ANI is simulated using the TechSim 2003. This shows the nanotechnology is very much affected by the NanoFab. In the final year 2020, the NanoFab and Presidential Change effects concerning simulation is 3.1 (= 47/15) times higher than the case of non-NanoFab and Presidential Change effects case. However, when just the Presidential Change effect is considered, the value is 1.2 (= 18/15). The RandD budget in 2002 was about 40 billion won (4 × 10^{11} won, 36.36 million US dollars). Considering the simulation, the RandD in 2020 is supposed to be about 124 billion won (3.1 × 40 billion won, 112.73 million US dollars) and 48 billion Won (1.2 × 40 billion won, 43.63 million US dollars) in each case. So, the NanoFab contributes very highly in the development of ANI. Therefore, the economic factor is much more important than the political factor.

4.6.3. Summary

The ANI is simulated which is not much different from the NNI. The focus of the ANI is the nuclear industry. The TechSim 2003 is introduced. The ANI promises that the conventional NPP oriented nuclear engineering has the competitive ability to the other industry, because the nanotechnology is fully applied to the NPPs research field.

4.7. GEN-4 AND GNEP COLLABORATION

The initiatives of the Gen-4 and GNEP are based on the development of the mid-20th century studies. However, the Light Water Reactor (LWR) has showed the successful industrialization. This includes the Pressurized Water Reactor (PWR), Boiling Water Reactor (BWR), and Heavy Water Reactor (HWR) as CANDU (CANada Deuterium Uranium). The initiatives have requested the high economical performances, the waste reductions, and the safety related non-proliferation. However, there is a significant limitation of the development due to the lower technology level. The nano-scale implementation could make the other solution. For example, the waste of the Spent Nuclear Fuels (SNFs) needs many years to be stabilized for the safe state. The NT can show new kinds of the ways to treat for reducing the toxicity as well as the quantity. As announced in the GNEP plan, the computation molecular simulation can show us the details of the material structure in the atomic level. This is the original meaning of the nano-scale research.

The technology can make the new method in not only the nuclear materials and waste form, but the thermal-hydraulics and nuclear reactor theory, which are the conventional and original nuclear research areas. The molecular scale simulation is used in the fluid simulation of the coolant which is called as the Computational Fluid Dynamics (CFD) in the nuclear thermal-hydraulics. It is also used in the nuclear reactor analysis as the parallel computing as the conjunction technology of NT and IT. This means another combination of the interdisciplinary technology of the NT and IT.

REFERENCES

[1] G.H. Miley, G. Narne, T. Woo, 'Use of Combined NAA and SIMS Analyses for Impurity Level Isotope Detection', Journal of Radioanalytical and Nuclear Chemistry, Vol. 263, pp. 691-696, N0.3, 2005.

[2] Y. Iwamura, M. Sakano, T. Itoh, 2002. Elemental Analysis of Pd Complexes: Effects of D2 Gas Permeation, Jpn. J. Appl. Vol. 41, 4642-4650.

[3] Y. Iwamura, T. Itoh, M. Sakano, S. Sakai, S. Kuribayashi, 2003. Low Energy Nuclear Transmutation in Condensed Matter Induced By D2 Gas Permeation Through Pd Complexes: Correlation Between Deuterium

Flux And Nuclear Products, the 10th International Conference on Cold Fusion, Cambridge, MA.
[4] Z. F. Ziegler, J. P. Biersack, U. Little mark, 2003. The Stopping and Range of Ions in Solids, Pergamon Press, New York.
[5] J. P. Biersack, Z. F. Ziegler, 1982. The Calculation of Ion Ranges in Solids with Analytic Solutions, 157-176, Ion Implantation Techniques, Springer Verlag, Berin, Germany
[6] Z. F. Ziegler, 1988. Ion Implantation Science and Technology, 2nd Ed., Academic Press, Inc.
[7] T. H. Woo, H. S. Cho, 'Nano-scopic measurement for radiation of nuclear waste forms using ion beam injection in the drum treatment', *Nuclear Instruments and Methods in Physics Research Section A: Accelerators, Spectrometers, Detectors and Associated Equipment*, 2011. (in press)
[8] J. H. Saling, A. W. Fentiman, 2002. Radioactive Waste Management, 2nd Ed., Taylor and Francis.
[9] Ulf Lindblom, Paul Gnirk, 1982. Nuclear Waste Disposal, Can We Rely on Bedrock?, Pergamon Press.
[10] J. F. Ziegler, J. P. Biersack, U. Little mark, 1985, 2003. The Stopping and Range of Ions in Solids, Pergamon Press, New York.
[11] J. P. Biersack, J. F. Ziegler, 1982. The Calculation of Ion Ranges in Solids with Analytic Solutions, 157-176, Ion Implantation Techniques, Springer Verlag, Berin, Germany.
[12] J. F. Ziegler, 1988. Ion Implantation Science and Technology, 2nd Ed., Academic Press, Inc.
[13] R. A. Johnson, A. N. Orlov, Physics of Radiation Effects in Crystals, North-Holland Physics Publishing, Elesevier Science Publishing Company, Inc. (1986).
[14] M. A. Kumakhov, F. F. Komarov, Energy Loss and Ion Ranges in Solids, Gordon and Breach Science Publishers (1979).
[15] Taeho Woo, Taewoo Kim, 'Design of thermal-hydraulic cooling in nano-scale environment of lunar nuclear power plant', Int. J. Nuclear Energy Science and Technology, Vol. 5, No. 4, 298-309, 2010.
[16] Potter AE, et al. Physics and astrophysics from a lunar base, first NASA workshop, Stanford, CA 1989.
[17] Heiken, et al. Lunar Sourcebook, a user's guide to the Moon. New York: Cambridge University Press 1991.
[18] Prado M. Introduction to PERMANENT, http:// www.permanent.com, General dynamics/Convair under contract to NASA, 2009.

[19] [19] Eckart P. The lunar base handbook : an introduction to lunar base design, development, and operations, New York, McGraw-Hill, 1999.
[20] [20] Informatics. The life of Konstantin Eduardovitch Tsiolkovsky, http://www.Informatics.org/ Museum/tsiol.html, Retrieved January 12, 2008.
[21] Altair. Rinehart's floating moonbase. 1959.
[22] Dept. of the Army. Project Horizon, A U.S. Army Study for the Establishment of a Lunar Military Outpost, I, Summary. 1959.
[23] Yonhap News Agency. Software glitch halts rocket launch. 2009.
[24] USNRC. SECY-93-092, PRA attachment 4. USNRC Commission Papers. 1993.
[25] Tae-Ho Woo, 'Management of energy policy in atomic-multinology (AM) using system dynamics (SD) method', Annals of Nuclear Energy, Vol. 37, Issue 5, 707-714, 2010.
[26] Tae Ho Woo, George H. Miley, et al, 'The abundant Excess Heat Production during Low Energy Nuclear Reaction in the Nano Scale Solid-The Cold Fusion, 14 years' legacy', Proceedings of the Korean Nuclear Society Spring Meeting, Kwangju, May 2002.
[27] Little Big Science, Gary Stix, Scientific American, Vol. 285, Number 3, September 2001.
[28] Nanotechnology Research Directions: IWGN Workshop Report Vision for Nanotechnology Research and Development in the Next Decade, WTEC, Loyola College in Maryland, September 1999.
[29] M. J. Schmidt, M. S. Gary, Combining Systems and Conjoint Analysis for Strategic Decision Making with an Automotive High-Tech SME, System Dynamics Review, Vol. 18, No. 3, 359-379, Fall 2002.
[30] M. Liehr, A. Grobler, M. Klein, P. M. Milling, Cycles in the Sky : Understanding and Managing Business Cycles in the Airline Market, System Dynamics Review, Vol. 17, No. 4, 311-332, Winter 2001.

PROBLEMS

1. What does the low energy nuclear reaction (LENR) stand for?

Solution

This LENR (Low Energy Nuclear Reactions) is well known as Cold Fusion. In addition, the CANR (Chemically Assisted Nuclear Reactions) is another term for this phenomenon. Cold fusion is *nuclear fusion* of atoms at conditions close to room temperature, which is the contrast condition against well-understood fusion reactions such as those inside stars and high energy experiments. After nuclear fusion was reported in a tabletop experiment involving *electrolysis* of *heavy water* on a *palladium* (Pd) electrode *by Martin Fleischmann*, then one of the world's leading *electro-chemists,* and *Stanley Pons* in 1989, the experiment has been famous. They reported anomalous heat production as excess heat of a magnitude where they asserted would defy explanation except in terms of nuclear processes. The measuring small amounts of nuclear reaction byproducts, including *neutrons* and *tritium*, were reported. A cheap and abundant source of energy has been challengeable by the LENR.

2. Using $E=MC^2$, calculate the amount of energy generated from converting four hydrogen atoms into one helium atom.
The hydrogen atom mass : 1.673×10^{-24} gm.
The helium atom mass : 6.645×10^{-24} gm.

Solution

The nuclear fusion process makes 4 H atom → 1 He atom.
The energy created per one nuclear fusion event is as follows;
$\Delta E = [4M_H C^2] - [M_{He} C^2]$
$\Delta E = [(4 \times 1.673 \times 10^{-24}$ gm$) \times (3.0 \times 10^{10}$ cm/s$)^2] - [(6.645 \times 10^{-24}$ gm$) \times (3.0 \times 10^{10}$ cm/s$)^2] = 4.23 \times 10^{-5}$ ergs

3. Explain the crystal lattice energy of LiI?

Solution

The lattice energy is the difference between energy of gaseous ions and solid crystal. So, Li(+)(g) + I(-)(g) → LiI(s).

The solution is ΔH (= crystal lattice energy).

Using detail equations,
Li(s) → Li(g)　　　　　$\Delta H = 155$ kJ/mole
Li(g) → Li(s)　　　　　$\Delta H = -155$ kJ/mole

Li(g) → Li(+)(g)   ΔH = 520 kJ/mole
Li(+)(g) → Li(g)   ΔH = -520 kJ/mole

$I_2(s)$ → $I_2(g)$ ΔH = 62 kJ/mole
1/2 $I_2(g)$ → 1/2 $I_2(s)$ ΔH = (-62/2) kJ/mole

I2 (s) → 2I(s)   ΔH = 151 kJ/mole
1/2 2I(g) → 1/2 $I_2(g)$ ΔH = (-151/2) kJ/mole
I(g) → I(-)(g)   ΔH = -295 kJ/mole
I(-)(g) → I(g)   ΔH = 295 kJ/mole
Li(s) + I(s) → LiI(s)   ΔH = -271 kJ/mole

Therefore,   Li(+)(g) + I(-)(g) → LiI(s)   ΔH = -757.5 kJ/mol

4. Using $dpa = (1 \times 10^{-16}) N_d \cdot D_c/\rho_n$, calculate critical dose. For $dpa = 5 \times 10^{-13}$, $N_d = 2.1$ in water.

5. Find the Nu and Pr numbers in zero or near-zero gravity?

Chapter 5

Conclusion

5.1. Abstract

In this book, the new century oriented technology has been introduced. It could affect to a significant results, although it is not an imminent usages in the power plant site as well as some business' places like the hospital or the financial business office. The IT, BT, and NT are not the conventional technologies. The application skills are under development. Maybe, it is used for our lives in the mid-21st century.

However, the important thing is that the necessity of the interdisciplinary research in the nuclear industry is deficient. The need is originated by the politician in the nuclear power related governments. The promotion of Gen-4 and GNEP is a good example that shows us that the wrong guide of the research could not be successful.

Although the research fund is not easily related to the commercialization, the vast investment should be done in the reasonable aspect, because it is from mostly the tax money. Some countries like South Korea have pushed up the research fund. The guide of the research direction is very important in order not to take the non-successful industrialization cases of the Gen-4 and GNEP.

The main object of the industrial development in the nuclear energy is to make us take a better life. It could be achieved with the accompanied technology in the general standard. The IT, BT, and NT are the general trends for the science and technology in this era.

Lastly, there is the simulation of the Atomic Multinology which shows the trend of the political progress in the aspect of the interested points [1].

5.2. THE SIMULATION OF THE ATOMIC MULTINOLOGY

5.2.1. Abstract

New kind of technology is promoted for the marketing creation. The 3 kinds of the technologies as the info-technology (IT), nano-technology (NT), and bio-technology (BT) are applied to the nuclear technology. A new field, Atomic-Multinology (AM) is initiated and modeled for the dynamic quantifications. The System Dynamics (SD) algorithm is used in the dynamical simulation for the management of the projects. There are 2 major models which include the Funds and the Academic Factor. The result shows that the successfulness of the AM increases, where the 100 months are the investigated period. The values of the dynamical simulation increase slowly in early stage and fast in later stage, which means that there is the time necessity to adapt to new technology field in the industry as well as the academic area.

5.2.2. Background

The interdisciplinary promotion in our lives is one of most important characteristics in the 21st century. The NT, IT, and BT are investigated for the interdisciplinary application to the nuclear technology. The examinations for the industry as well as academic field are useful to make the management for the new field of the technology. Globally, 436 nuclear power plants (NPPs) are operating in 2009 which is seen in Table 5.1. So, these NPPs are the object of the application using the interdisciplinary technologies. The conventional nuclear technology has focused on the NPPs related areas. There are several comparisons of the academic areas between the conventional and AM classifications in Table 5.2 where the nuclear energy aspect is considered. The new trend of the academic progress is shown in Figure 5.1. The AM is expected to be developed like this procedure in the initial stage of the academic aspect. Figure 5.2 shows the combination of 3 kinds of different technologies.

The NT was initiated for the new technology innovation in 1990s as the National Nanotechnology Initiative (NNI) which has promoted to be applied to the wide ranges of science and technology. The major object is to make the better efficiency in the variety of fields. The nanotechnology application to the nuclear technology has been introduced to the nuclear society. This promotion highlights the better solution of the stagnated marketing in the NPPs' construction situation. Considering historical aspect, Albert Einstein calculated the size of a single sugar molecular from experimental data on the diffusion of sugar in water. The work showed that each molecule measures about a nanometer in diameter [1]. Around 1960s, Richard Feynman mentioned there were plenty of rooms at the bottom of the matter. Later in 1990s, the nano-scale research was initiated nationally by several governmental members in the United States. The NNI plan of the National Science and Technology Council (NSTC) went to President Clinton, and, subsequently, the budget was submitted to congress in 1999 which was referred as the NNI officially [3, 4, 5]. The goals listed in the NNI documentation [6] are to advance a world-class nanotechnology research programs and to develop educational resources, a skilled workforce, and the supporting infrastructure tools. Hence, the NNI has been led by the government for the political purpose, which deeply depends on the funds of the nation.

Table 5.1. World Nuclear Power Generation Units (Nuclear Energy Institute, August 2009)

Country	Number of unit	Country	Number of unit
Argentina	2	Mexico	2
Armenia	1	Netherlands	1
Belgium	7	Pakistan	2
Brazil	2	Romania	2
Bulgaria	2	Russia	31
Canada	18	Slovakia	4
China	11	Slovenia	1
Czech RP	6	South Africa	2
Finland	4	Spain	8
France	59	Sweden	10
Germany	17	Switzerland	5
Hungary	4	Taiwan, China	6
India	17	U.K.	19
Japan	53	U.S.	104
Korea Rep.	20	Ukraine	15
Lithuania	1	Total	436

Table 5.2 Comparisons of classification in nuclear technology

Conventional	Atomic multinology (AM)
Nuclear reactor theory	Atomic info-technology (AI)
Nuclear safety analysis	
Nuclear materials	Atomic nano-technology (AN)
Nuclear chemistry	
Nuclear thermohydraulics	
Radiation detection	Atomic bio-technology (AB)
Radiation biology and medicine	

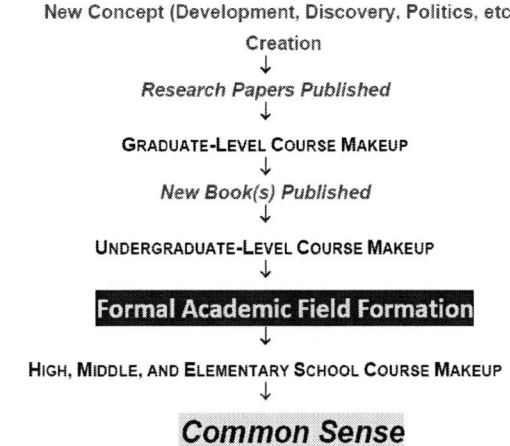

Figure 5.1. Procedure of the new academic field.

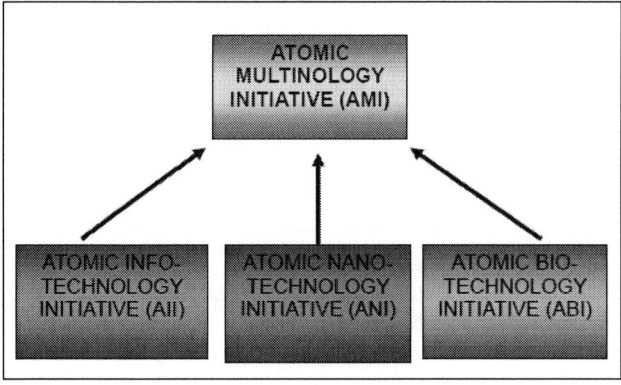

Figure 5.2 Strategy for the Atomic Multinology Initiative(AMI).

The IT has been represented by the computer and its applications. Although the first use of the word 'computer' was recorded in 1613, the word began to take on its more familiar meaning, describing a machine that carries out computations [7]. The father of the modern digital computer is recognized by George Stibitz in 1937 [8]. Currently, the IT has encompassed a variety of aspects of computing and technology, which is to transmit, convert, process, store, and protect. In this chapter, it focuses on the nuclear industry with the NPPs.

The BT has been developed in the areas of the biology, medicine, agriculture, and food science. The United Nation defined any technological application that uses biological systems, dead organisms, or derivatives thereof, to make or modify products or processes for specific use [9]. Even though the application of the BT has been used from B.C. 200, the field of modern biotechnology is thought to have largely begun on 1980, when the United States ruled that a genetically-modified micro-organism could be patented in the case of Diamond Chakrabarty [10]. Biological sciences are called commonly as the life sciences in the industrial applications including nuclear technology.

Current global nuclear industry has the stagnation in promoting the NPP constructions due to the long-period continued anti-nuclear mood. Although the Gen-4 and global nuclear energy partnership (GNEP) have been promoted in the several countries which has been led by the United States, the substantial result has not been come out yet. So, new spirit of the industry boosting in the nuclear community is urgently needed. The conventional classification of the nuclear industry has the limitations to be adapted in the 21st century style technology where the IT, NT, and BT are applied in the broad wide industries. Therefore, it is necessary for the classical nuclear industry to make use of this new technology concept. Using non-linear method like the SD, the energy policy is suggested. The non-linearity of the simulation can make the future expectation comparatively easily, because the linear algorithm can show just the exact feature where the uncertainty of the future event cannot be seen well. The meaning of the uncertainty of SD is explained as nonlinearity, stock-flow, feedback and time paths. The basic concept of the hypotheses is shown Table 5.3.

Conventional energy policy promotions like the Gen-4 and GNEP has not been successful until now, especially in the United States. Therefore, it is very important for the nuclear communities to make the computational simulation for the possible energy policy before making any progress of the energy research and development promotion. According to this reason, the SD has many previous experiences during last 50 years. There are a policy for the industrial promotion [11], a policy for the civic reformation in the Boston city [12], and a simulation for the world organizations [13]. Recently, there are some decision-making related papers [14, 15]. Therefore, SD is a suitable algorithm in the management of the energy policy, which has the numerical verification method. The numerical expression can enhance the reliability of the decision-making result, because the comparisons of the dynamic quantification can show the priority factor of several outputs.

Table 5.3. Basic concept of hypotheses

Classification	Hypotheses
Motivation	Marketing promotion of the NPPs' industry
	Stagnation of Gen-4 and GNEP initiatives
Major method	Nuclear application using info-technology (IT), nano-technology (NT) and bio-technology (BT)
	New classification from the conventional academic sorting system
Simulation tool	System dynamics (SD)
Final goal	Advanced energy policy construction

The simulations using SD are constructed for the time dependent quantification of the management of the AM. The chapter 5.2.3 explains the method of the research. The calculation for the modeling is shown in the chapter 5.2.4. The chapter 5.2.5 describes results of the research. There are some summaries in the chapter 5.2.6.

5.2.3. Method

The SD was created by Dr. J. Forrest in Massachusetts Institute of Technology (MIT) for the quantifications of the systematic situations. The applications for the non-linear characteristics of the social and economical system have been studied. For the quantification, it is to test and model the complex features in the dynamical scenarios of the interested matters. The Figure 5.3 shows the fundamental algorithm of the problem solving with SD. There is the paralleled configuration between the real world and the systems thinking world which shows the characteristics of the construction in the modeling. It has been meant that the systems thinking is any process to problem solving, as viewing 'problems' as parts of an overall system, rather than reacting outcomes or events and potentially contribution to further development of the undesired issue and problem. Therefore, the SD is the dynamical algorithm of the systems thinking. It has been published for the organizations by the transitions of the time [11, 12, 13, 16]. In addition, there are some decision-making related papers [14, 15, 17]. There are the dynamic simulation methods using the SD where some software as the Vensim [18], Powersim [19], and ITHINK [20] were applied for the quantifications. In this chapter, the Vensim is used for the simulations.

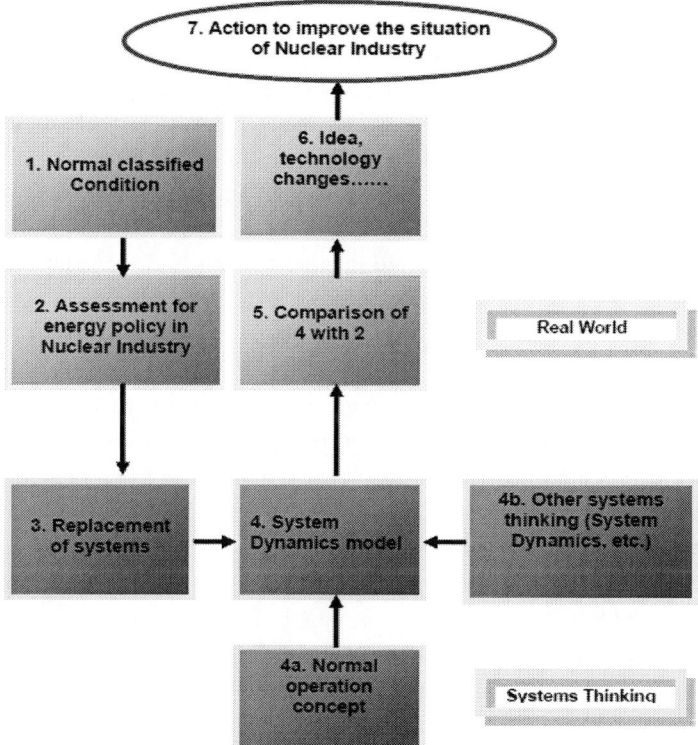

Figure 5.3. Real World vs. Systems Thinking World.

The characteristics of the SD are a powerful methodology and computer simulation modeling technique for framing, understanding, and discussing complex issues and problems, which is explained by M. Radzicki in this chapter[21]. Namely, when the SD was developed, it was used to help managers improve their understanding of industrial processes, which is currently being used throughout all kinds of policy analysis and design. Historically, at the end of World War II, J. Forrester focused on the creation of an aircraft flight simulator for the U.S. Navy. After the digital aircraft simulator production, it was applied to the testing of computerized combat information systems. The MIT Digital Computer Laboratory was founded and placed under the direction of J. Forrester in 1947. The WHIRLWIND I was created which was MIT's first general-purpose digital computer and the digital computers could be effectively used for the control of combat information systems. After this project, J. Forrester got a leader of a division of MIT's Lincoln Laboratory in his efforts to create computers for the North American SAGE (Semi-Automatic Ground Environment) air defense system. The computers created by Forrester's team during the SAGE project were installed in the late 1950s, remained in service for approximately twenty-five years. SD provides the basic building blocks which can construct models of how and why complex real-world systems behave the way they do over time. The goal is to leverage this added understanding to design and implement more efficient and effective policies. It is necessary to understand the dynamic behavior of a system, its key physical and information stocks, flows and feedback structures for SD. There are several specialty of the SD as follows which were explained briefly in Chapter 2.3.

Nonlinearity: A large part of the SD modeling process involves the application of common sense to dynamic problems. A good SD operator should always see model behaviors that do not make sense. Such behaviors usually indicate a nonlinearity of the events. This is seen as single and double arrow lines in the modeling. Namely, the arrow line shows the event flow without any restriction.

Stock-flow: In SD modeling, it is the principle of accumulation to be raised by dynamic behavior. This means that all kinds of dynamic behaviors could be happened when flows accumulate in stocks, which is seen as EXAMPLE for accumulation and INPUT/PUTPUT for flows in Figure 5.4. It is like a bathtub where a flow can be thought of as a faucet and pipe assembly that fills or drains the stock. It is considered as the simplest dynamical system in the stock-flow structure. In SD, both informational and non-informational object can move through flows and accumulate in stocks.

Feedback: The stocks and flows in real world systems are part of feedback loops. The feedback loops are often joined together by nonlinear couplings where any object often cause counterintuitive behavior. This is seen as feedback loop in Figure 5.4. The plus sign means for the addition to

EXAMPLE of the feedback value, OUTPUT. Otherwise, if the sign is minus, the feedback value, OUTPUT, is subtracted from the EXAMPLE.

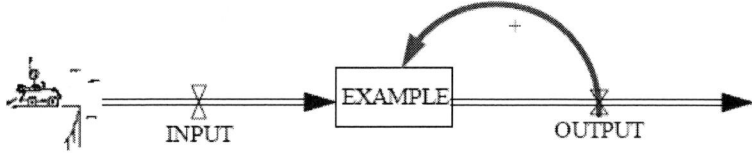

Figure 5.4 Stock- flow and Feedback.

Time paths: The most important thing of the SD modeling is the dynamic behavior of systems, where the operator tries to identify the patterns of behavior exhibited by interested system variables, and then builds a model with the characteristics of patterns. If a model has this capability, it can be used as a laboratory for testing policies aimed at altering a system's behavior in desired ways. This is seen as the single and double arrow lines. The lines mean event flows as well as time flows.

5.2.4. Calculation

The object of this chapter is to find out the trend of the successful management in the new technology field. The period is 100 months. So, the simulation shows the dynamic transition for the successfulness of the AM during the interested time as the month unit. There is the main modeling of the dynamic simulator for the AM in Figure 5.5. The connections with 3 different technologies are described by the dotted lines. There are 2 major sub-cases as the Funds and the Academic Factor, which means that the industrial affect is as the Funds and the research affect is as the Academic factor. The industrial and research affects represent the total contributions to the new R&D area for marketing management, especially to the AM incorporated with nuclear industry. The arrow line shows the direction of the incident (including the dotted line). Figure 5.6 shows SD modeling for Funds where Technological, Economical, and Political Factors are included. Each factor has the Innovation, Technological Progress, Investment, Tax, Party Strategy, and President Plan. Figure 5.7 shows SD modeling for Academic Factor where Study Interest, Test, Education Material, and Lecture Ability are included. Especially, the Test Effect has the feedback algorithm in Figure 5.8. The periodic affect is simulated by the feedback algorithm of SD. The quantifications are done by the Monte-Carlo method in each event. The test in Figure 5.9 shows the periodic change of 5 months which is seen by the 2 lines. The double arrow lines in Figure 5.8 shows the logical event flow like the cumulative values which is incorporated with feedback logic and periodic quantification.

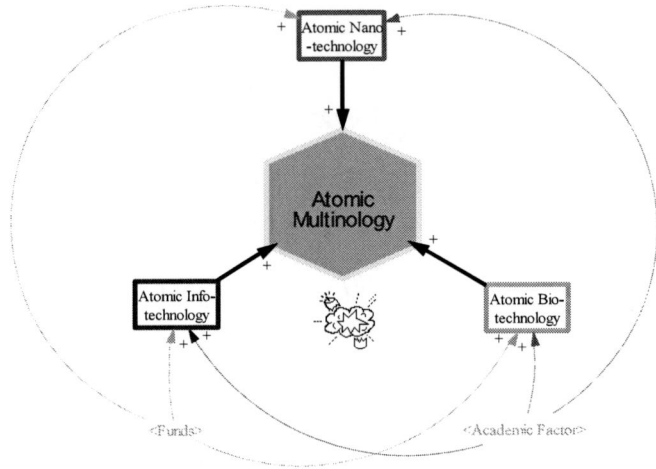

Figure 5.5. System dynamics (SD) modeling for Atomic Multinology (AM).

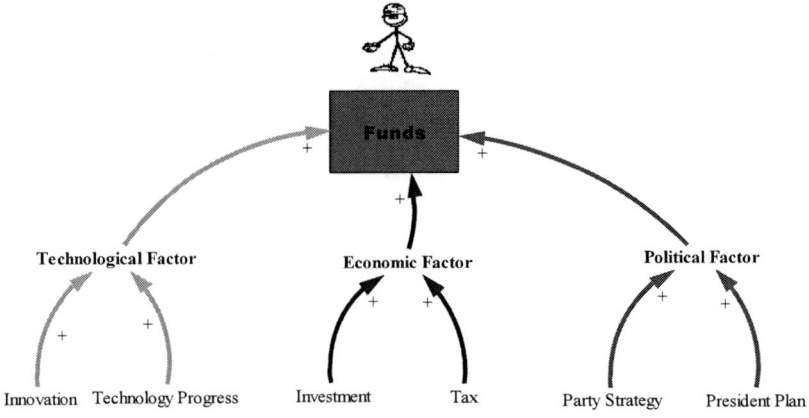

Figure 5.6 System dynamics (SD) modeling for Funds.

Conclusion 107

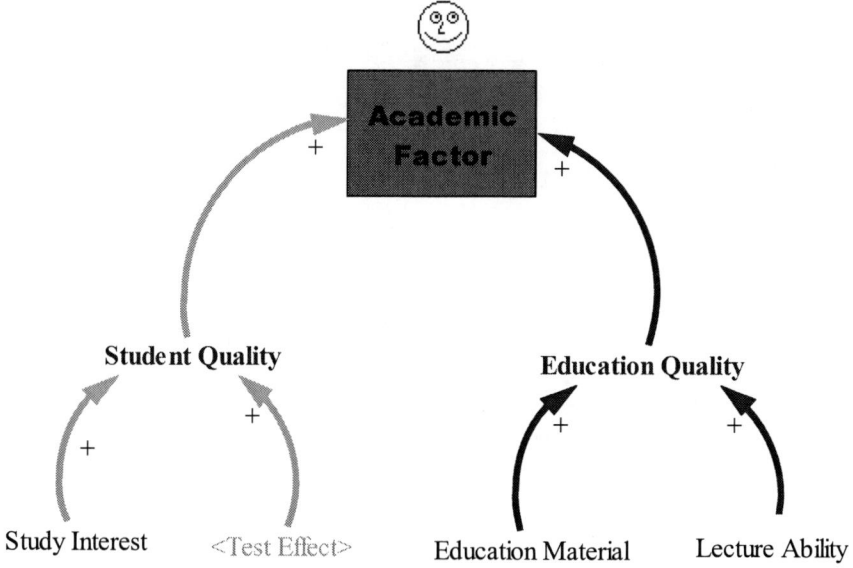

Figure 5.7. System dynamics (SD) modeling for Academic Factor.

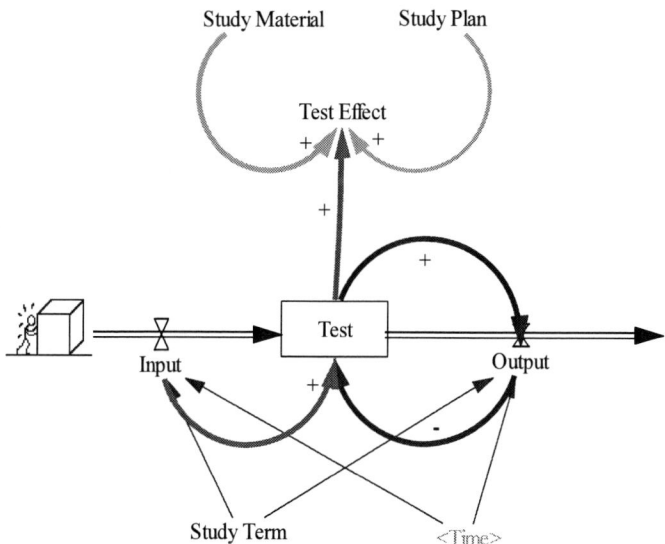

Figure 5.8. System dynamics (SD) modeling for Test Effect.

108   Taeho Woo

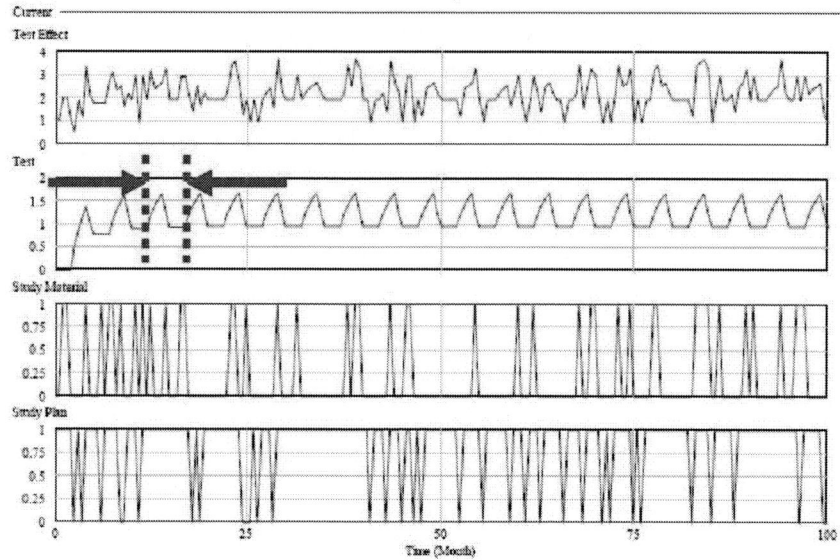

Figure 5.9. Simulation of Test.

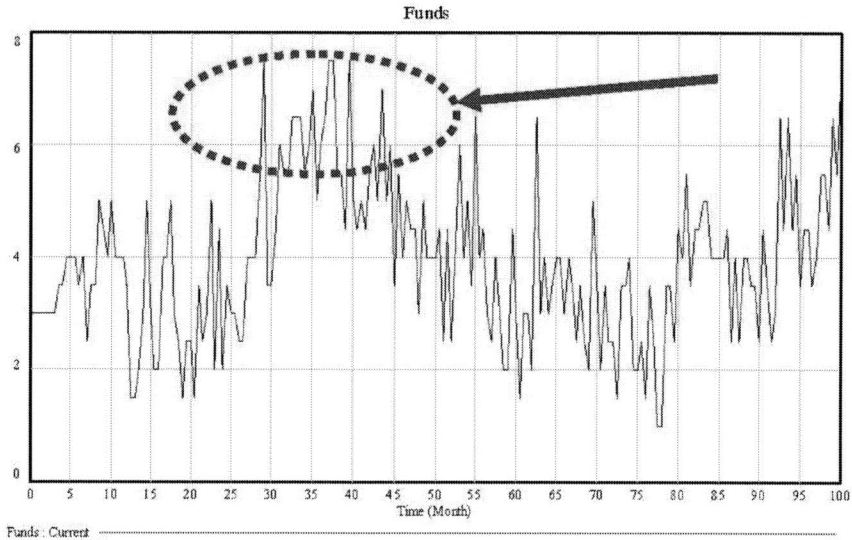

Figure 5.10. Simulation of Funds.

5.2.5. Result

There is the dynamic simulation of the funds in Figure 5.10 which is the relative value. In the simulation, the numerical values mean relative values. Namely, in case of the Funds which is one of two major factors in AM, the difference of the highest and lowest values is the characteristics of the variable. So, the meaning of the difference is the money. The value is 7.5 for the highest values from 29th month to 39th month. The value is 1.0 for the lowest values on 78th month. Therefore, the difference is 6.5 which is the relative value of the money. The relative value of the funds is the currency in the modeling country like the U.S. dollar. Otherwise, in case of Academic Factor, the value means the importance which is the characteristics of the variable. That is to say, the importance of the Academic Factor increases, which is composed of the Student Quality and the Education Quality. The Figure 5.11 shows the values of Academic Factor which increase nearly linearly. The Study Interest, Test Effect, Education Material, and Lecture Ability increase as the summation value. All quantifications of these 4 variables are done as the Monte-Carlo calculation of the random sampling. Using the simulation of the AM which are calculated by the Atomic Info-technology, Atomic Nano-technology, and Atomic Bio-technology, the relative value are obtained. This value means the degree of successfulness in AM. So, as the value is higher, the possibility of the success in AM increases. Although the meaning of 'success in AM' cannot be defined easily, one can understand 'successfulness of AM' as the comparative values which are changed by the dynamical manner. Table 5.4 shows the classifications of the numerical values in the graphs. The final result of the AM is shown in Figure 5.12. The values increase slowly in early stage and fast in later stage. The values are normalized where the 1.0 is the highest value of the simulation. The initial value is 0.00. The highest value is 1.00 in 100th month. The median value is seen around 70th month, which means the values increase slowly.

As Table 5.3 is noted, the final goal of this chapter is to make an advanced energy policy. The major factors are applications of the IT, NT, and BT into the nuclear industry. Then, the result gives relative numerical values as the successfulness of AM in energy policy.

As the classification of the simulation is seen in Table 5.4, the Funds and Academic Factors are 2 major points for the AM in this chapter. The 'Funds' means the 'development' and the 'Academic factor' means the 'research'. So, these 2 terminology of 'Funds' and 'Academic factor' represent the whole characteristics in the stage of the new industrial construction, because the

R&D is called as a representative terminology of any industrial field like in the fields of the automobile industry of General Motors and Hyundai, the content industry of Sony Music and Warner Brothers Movie, and the airplane industry of Boeing and Airbus.

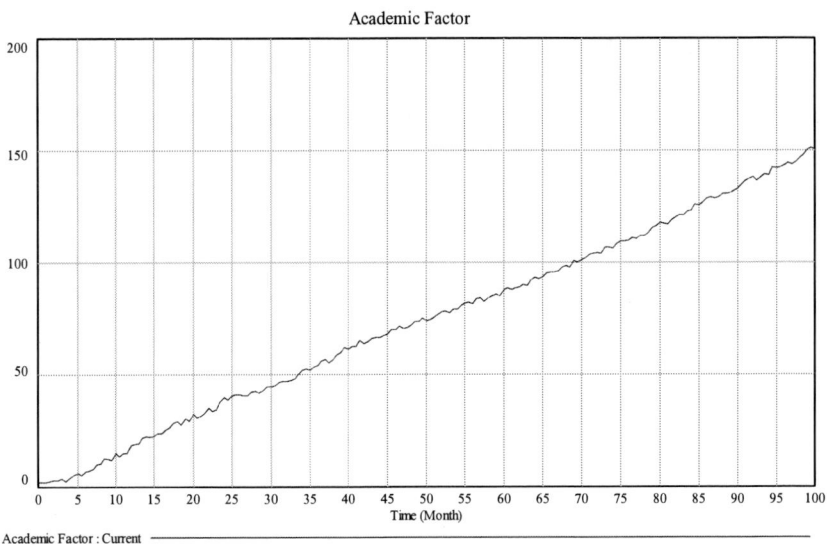

Figure 5.11. Simulation of Academic Factor.

Table 5.4. Classifications of the numerical values in the graphs

Simulation	Numerical meaning
Funds	Money
Academic factor	Importance
Atomic multinology (AM)	Successfulness

The dynamic simulation is performed for the AM. The successfulness of the AM increases slowly and falls down fast, which could be the fund or the object like the research project papers. This is due to the time necessity to be adapted to the new technology field. Although the values of the Funds increase rapidly in early stage, the Academic Factor has the higher values. So, the effect of the Funds is negligible. The new industrial field has the academic characteristic in early steps because there are many kinds of the researches before the commercialization of the specified products. This trend of the graph could be expressed as other technology promotion in the early stage. The following are some conclusions of this chapter.

- The successfulness of AM is described well by the SD of the time step and feedback algorithm.
- The graphical and colorful configurations are used for the meaning of the incident.
- The single and double arrow lines mean the direction and logical expression of the event flows respectively.
- The new creation of the industrial field is simulated by the dynamical quantification.
- The modeling is used successfully to NPPs.

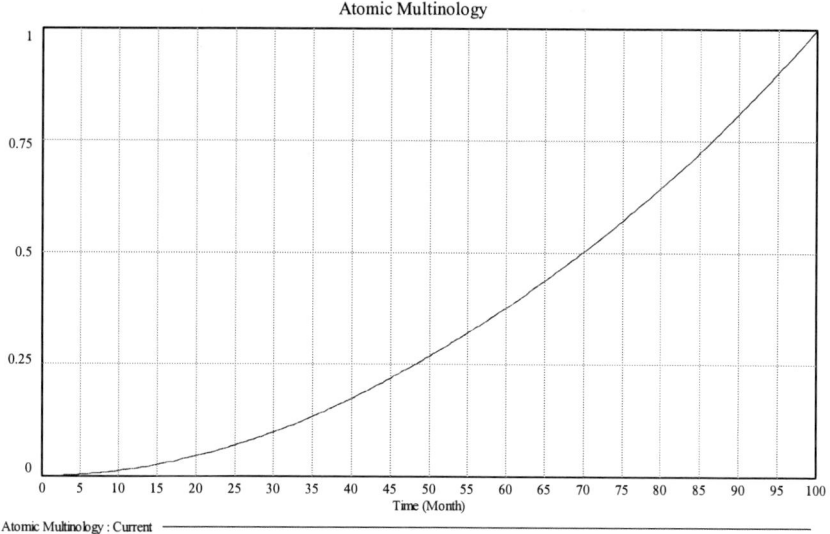

Figure 5.12. Simulation of Atomic Multinology (AM).

For future research, it is needed to consider some more investigations for comparisons of several nations which have different situation of the nuclear industry. In case of the Republic of Korea, the construction of NPPs is very active. So, it is easy to model the new technology promotion in the nuclear industry.

5.2.6. Summary

The dynamic simulation for the AM shows the another kind of technology innovation where the successfulness of the AM increases slowly and falls down fast, which could be the fund or the object like the research project papers. It is applicable to make a model the new technology promotion in the nuclear industry.

5.2.7. Caption

Table 5.1 World nuclear power generation units (Nuclear Energy Institute, August 2009).
This shows the number of unit for the operating nuclear power plants.

Table 5.2 Comparisons of classification in nuclear technology.
This shows the comparisons of the classification between the conventional and newly introduced Atomic Multinology (AM) ways.

Table 5.3 Basic concept of hypotheses.
This shows the basic concept of hypotheses.

Table 5.4 Classifications of the numerical values in the graphs.
This shows the classifications of the numerical values in the graphs.

Figure 5.1 Procedure of the new academic field.
The procedure of the academic progress is presented for the new academic creations. This is the configuration for the formalization of the important steps. The research paper is published firstly. Then, it is developed to the common sense stage.

Figure 5.2 Strategy for the Atomic Multinology Initiative (AMI).
Newly introduced Atomic Multinology Initiative (AMI) is composed of 3 kinds of contents as the initiatives of info-technology (IT), nano-technology (NT), and bio-technology (BT). This is a different viewpoint classification from the conventional classification where the Nuclear reactor theory, Nuclear safety analysis, Nuclear materials, Nuclear chemistry, Nuclear thermohydraulics, Radiation detection, and Radiation biology and medicine are the major research areas.

Figure 5.3 Real World vs. Systems Thinking World.
This shows the comparison between real world and systems thinking world. The real world means the situations of the event happening and the event progressing. Otherwise, the systems thinking world means the situations which is a base of the system dynamics manipulations. The systems thinking is any process to problem solving, as viewing 'problems' as parts of an overall system, rather than reacting outcomes or events and potentially contribution to further development of the undesired issue and problem.

Figure 5.4 Stock-flow and Feedback.
This is the configuration of the Stock-flow and Feedback.

Figure 5.5 System dynamics (SD) modeling for Atomic Multinology (AM).
This is the configuration of the AM where the arrow line and dotted arrow line mean the event flow. Each arrow shows the plus sign or minus sign. The boxes are the event name which is drawn differently for the visual recognition. The values of each event are done for the judgment as the random sampling or the constant value. The parenthesis is another event flow. Namely, <Funds> is connected to Figure 5.6 System dynamics (SD) modeling for Funds. <Academic Factor> is connected to Figure 5.7 System dynamics (SD) modeling for Academic Factor.

Figure 5.6 System dynamics (SD) modeling for Funds.
The top event, Funds is composed of Technological Factor, Economical Factor, and Political Factor. The Technological Factor is composed of Innovation and Technology Progress. The economic factor is composed of Investment and Tax. The Political Factor is composed of Party Strategy and President Plan. The non-parenthesis event also means the event which is quantified by judgment as the random sampling or constant.

Figure 5.7 System dynamics (SD) modeling for Academic Factor.
The top event, Academic Factor is composed of student quality and education quality. The Student Quality is composed of Study Interest and Test Effect. The Education Quality is composed of Education Material and Lecture Ability. The non-parenthesis event also shows the event which is quantified by judgment as the random sampling or constant.

Figure 5.8 System dynamics (SD) modeling for Test Effect.
The top event, Test Effect is composed of Rest, Study Material, and Study Plan. The Test means the logical rate of the Input and Output. The arrows show the event flow with the plus or minus sign. In the figure, the function of Vensim code, QUNATUM (A, B) is used where the A is quantized by B. QUANTUM returns the number smaller than or equal to A that is an integer multiple of B (B * integer part of (A/B)). A common use of QUANTUM is to remove the non-integer part of a value (e.g., QUANTUM (3.456, 1.0) is equal to 3.0). If B is less than or equal to zero, then A is returned.

Figure 5.9 Simulation of Test.
This is the result of Test in Figure 5.8, where it shows the periodic change of 5 months which is seen by the 2 line arrow using QUANTUM.

Figure 5.10 Simulation of Funds.
This is the result of Funds in Figure 5.6, where it shows the highest values are from 29th month to 39th month as the dotted circle. The lowest value is shown in 78th month.

Figure 5.11 Simulation of Academic Factor.
This is the result of Academic Factor in Figure 5.7, where it shows the values increase nearly linearly.

Figure 5.12 Simulation of Atomic Multinology (AM).
This is the result of Atomic Multinology (AM) in Figure 5.5, where it shows the values increase slowly in early stage and fast in later stage.

REFERENCES

[1] Tae-Ho Woo, 'Management of energy policy in atomic-multinology (AM) using system dynamics (SD) method', Annals of Nuclear Energy, Vol. 37, Issue 5, 707-714, 2010.
[2] Stix, G., 2001. 'Little Big Science', Scientific American, Vol. 285.
[3] National Science and Technology Council (NSTC), 2009. National Nanotechnology Initiative, Research and Development Leading to a Revolution in Technology and Industry.

[4] National Science and Technology Council (NSTC), 2008. National Nanotechnology Initiative, Strategy for Nanotechnology-Related Environmental, Health, and Safety Research.
[5] WTEC, 1999. Nanotechnology Research Directions: IWGN Workshop Report Vision for Nanotechnology Research and Development in the Next Decade, WTEC, Loyola College in Maryland.
[6] National Nanotechnology Coordination Office (NNCO), 2009. Website of the NNI. *http://www.nano.gov/*.
[7] Oxford University Press, 1989. Oxford English Dictionary.
[8] National Inventors Hall of Fame Foundation, 2007. Inventor profile for George R. Stibitz.
[9] United Nation, 1992. *The Convention on Biological Diversity* (Article2. Use of Terms).
[10] Chakrabarty, *Diamond, 1980. 447 U.S. 303. No. 79-139. United States Supreme Court.*
[11] Forrester, J.W., 1961. Industrial Dynamics. Productivity press.
[12] Forrester, J.W., 1969. Urban Dynamics. Pegasus Communications.
[13] Forrester, J.W., 1971. World Dynamics. Wright-Allen Press.
[14] Kampmann, C.E., 1996. Feedback loop gains and system behavior. Proceedings of the1996 International System Dynamics Conference Boston. System Dynamics Society, Albany, NY, pp. 260–263.
[15] Liehr, M., Grobler, A., Klein, M., Milling, P.M., 2001. Cycles in the Sky : Understanding and Managing Business Cycles in the Airline Market, System Dynamics Review 17(4) 311-332.
[16] Forrester, J.W., 1968. Principles of Systems, 2nd Ed. Pegasus Communications.
[17] Forrester, J.W., 1975. Collected Papers of Jay W. Forrester. Pegasus Communications.
[18] Vensim, Ventana Systems, Inc.
[19] PowerSim, Powersim Software.
[20] ITHINK Software, ISEE Systems, Inc.
[21] Radzicki, M., Taylor, R., 1997. U.S. Department of Energy's Introduction to System Dynamics, A Systems Approach to Understanding Complex Policy Issues, Ver. 1.

ACKNOWLEDGMENT

The authors are thankful for the permission of publication with full acknowledgement of its original publication in the journal of Elsevier B.V.

In addition, acknowledgement is given to the journal of the Inderscience Publishers Ltd. (www.inderscience.com) in which it is published as the original source of publication.

INDEX

A

A scanning tunneling microscope (STM) 6
Abalone (Molecular dynamics, visualization) 7
ACEMD (Molecular dynamics) 7
AMBER (Molecular dynamics) 7
Atomic force microscopy (AFM) 6
Atomic multinology 2, 3, 93, 97, 98, 100, 106, 110, 111, 112, 113, 114

B

Biological technology (BT) 1, 33
BOSS (Biochemical and Organic Simulation System) 7
BRAHMS (Molecular dynamics) 7

C

CASTEP (Density-functional theory) 7
CCP5 (Program Library, various) 7
CHARMM (Chemistry at HARvard Molecular Mechanics) 7
CPMD (Molecular dynamics) 7

D

Dalton (Computational chemistry) 7
DiMol2D (Molecular dynamics) 7
DL_MESO (Dissipative particle dynamics) 7
DL_POLY (Molecular dynamics) 7
DYNAMO (Molecular dynamics) 7

E

EGO VIII (Molecular dynamics) 7
Electron Gamma Shower (EGS) 4, 9
ENCAD (Molecular dynamics) 7
Environmental Protection Agency (EPA) 2
ESPResSo (Molecular dynamics) 7

F

FOCUS (Molecular dynamics) 7

G

Gaussian (Electronic, Computational chemistry) 7
gdpc (Molecular dynamics visualization) 7
Generation 4 (Gen-4) 1, 8, 91, 97, 101, 102
Gerd Binnig and Heinrich Rohrer 6
GROMACS (Molecular dynamics) 7
GROMOS (Molecular dynamics) 7

Index

H
HOOMD (Molecular dynamics) 7

I
IMD (Molecular dynamics) 7
Information technology (IT) 1, 11

J
Jmol (Visualization) 7

L
Lord Ernest Rutherford 6
Los Alamos National Laboratory (LANL) 4
Low Energy Electron Microscopy (LEEM) 5

M
Monte-Carlo (MC) 2, 4
Monte Carlo N-Particle eXpanded (MCNPX) 4
Monte Carlo N-Particle Transport (MCNP) 4

N
Nano-scale technology 1
Nano technology (NT) 61
National Institute of Health (NIH) 2
Neutron Reflection Analysis (NRA) 6
Nuclear power plant(s) (NPP(s)) vii, 1, 2, 14, 17, 18, 20, 25, 61, 74, 76, 81, 86, 90, 92, 98, 101, 102, 111, 112

Q
QMGA (Visualization) 7

R
RasMol (Visualization) 7
RedMD (Molecular dynamics) 7
Rutherford Backscattering (RBS) 4, 6, 9

S
SageMD (Simulation front and back end) 7
Scanning Electron Microscope (SEM) 4
Secondary Ion Mass Spectroscopy (SIMS) 5
SIESTA (Molecular dynamics) 7
SMMP (Monte Carlo simulation) 7
SYBYL (Various) 7
System Dynamics (SD) 4, 14, 15, 16, 17, 20, 25, 26, 30, 31, 86, 88, 93, 98, 101, 102, 103, 104, 105, 106, 107, 111, 113, 114, 115

T
Tesla Bio Workbench (GPU computing) 7
Three Miles Island (TMI) 1
TINKER (Software tools for molecular design) 7
Transmission electron microscopy (TEM) 5

U
UHBD (Brownian dynamics) 7
Ultra high vacuum (UHV) 6

V
VASP (Molecular dynamics) 7
VMD (Visualization, Molecular dynamics visualization in 3-dimensions) 8

W

WIEN2K (Electronic structure calculation in solids) 8

X

X-PLOR (Computational structural biology) 8
X-ray photoelectron spectroscopy (XPS) 5

XCrysDen (Visualization, Crystalline and molecular structure visualization) 8

Y

YASARA (Fee and commercial) 8
YASP (Molecular dynamics) 8